BUSINESS MATH

David J. Hyslop
Bowling Green State University
Bowling Green, Ohio

Dennis Mathern
University of Findlay
Findlay, Ohio

GLENCOE
McGraw-Hill

New York, New York
Columbus, Ohio
Woodland Hills, California
Peoria, Illinois

Authors: David J. Hyslop is Chairman of the Department of Business Education at Bowling Green State University in Bowling Green, Ohio. He has also taught at Michigan State University and California State University at Los Angeles. Mr. Hyslop has written many articles and has taken an active role in numerous professional organizations, including that of past president of North Central Business Education Association.

Dennis Mathern is a business education instructor at University of Findlay in Findlay, Ohio. He has taught courses in business mathematics and business communications and has supervised student teachers. In addition, Mr. Mathern also has extensive business experience in the banking and finance industry.

Cover Design: Quarasan

Imprint 1997
Copyright © 1990 by Glencoe/McGraw-Hill. All rights reserved. Originally copyrighted in 1990 by Houghton Mifflin Company. All rights reserved. Printed in the United States of America. No part of this work may be reproduced or transmitted in any form or by any means, electronic or mechanical, including photocopying and recording, or by any information storage or retrieval system without the prior written permission from the publisher unless such copying is expressly permitted by federal copyright law. Send all inquiries to: Glencoe/McGraw-Hill, 936 Eastwind Drive, Westerville, Ohio 43081

ISBN: 0-395-44665-1

8 9 10 11 12 13 14 15 066 03 02 01 00 99 98 97

Contents

Pretest

Directions: Follow the instructions for each set of problems.

Solve the following problems.

1. $\begin{array}{r} 1.1 \\ +\ .002 \\ \hline 1.102 \end{array}$

2. $\begin{array}{r} {}^{412.91} \\ 653.05 \\ -327.19 \\ \hline 325.86 \end{array}$

3. $10.004 \times 1.50 = \underline{15.006}$

4. $835\overline{)15,030}$ ^{18}

Change the following written amounts to figures.

5. One thousand, twenty-eight __1,028__

6. Two million, eleven thousand, nine hundred sixty-three __2,011,963__

Round each number to the underlined place.

7. 4̲6 __50__

8. 2̲03 __200__

9. 98̲,471 __100,000__

10. 99̲8 __1,000__

11. 1.0̲6 __1.00__

12. 41.3̲9 __41.40__

13. 8.1̲42 __8.100__

14. .2̲06 __.200__

Reduce these fractions to their lowest terms.

15. $\frac{48}{54}$ __$\frac{8}{9}$__

16. $\frac{12}{84}$ __$\frac{1}{7}$__

17. $\frac{42}{48}$ __$\frac{7}{8}$__

18. $\frac{132}{11}$ __$\frac{12}{1}$__

Solve the following problems. Reduce answers to lowest terms.

19. $\frac{1}{3} + \frac{2}{9} =$ __$\frac{4}{3}$__

20. $12\frac{1}{4} - 7\frac{3}{8} =$ __$4\frac{6}{1}$__

21. $2\frac{3}{4} \times 1\frac{3}{8} =$ __$3\frac{25}{32}$__

22. $\frac{7}{12} \div \frac{1}{3} =$ __$\frac{21}{12}$__

Solve the following word problems.

23. During the summer Melissa earned money selling cosmetics door-to-door. In June she earned $782.15, in July $694.50, and in August $985.90. What were her total earnings for the summer?

    __$2,462.55__

24. Christopher scored 80% on his math exam. If there were 120 questions on the exam, how many did he answer correctly?

    __96__

25. Sarah made 15 free throws in 23 attempts. What was the percent of free throws she made?

    __1.5%__

26. A dozen eggs are selling for $1.40 this week, up $.20 from last week. By what percent did the price increase from last week?

_____

27. Barry got a loan of $1,500. He has to repay it in 9 months and will be charged 8% simple interest. What is the amount of interest he will pay?

_____

28. Hal's Appliances gives customers a cash discount of 2/10, n/30. If a customer receives a bill dated May 12 for $300, what is the amount due if the bill is paid on May 21?

_____

29. Wearever Shoe Store marks up its shoes 30% of cost. If a pair of shoes costs the store $25, what is the selling price?

_____

30. All salespeople at The Broadway Clothing Store are paid a weekly salary of $300 plus a commission of 10% on all sales over $10,000. If Louis Arund had $12,000 in sales for a week, what would be his gross pay for the week?

_____

31. Matthew is paid $8 an hour for a 40-hour week. All overtime work is paid at time-and-a-half. How much will he earn for working 60 hours?

_____

32. During February, Crystal wrote 6 checks totaling $482.26. She had a beginning balance of $1,022.46 and she deposited $156.00. What was her ending balance?

_____

33. Stacy's bank statement balance is $823.41. Stacy has two outstanding checks for $32.63 and $115.95. She also has an outstanding deposit of $416.83. What is her adjusted bank statement balance?

_____

34. James bought a used car for $3,500. He made a down payment of $500 and agreed to pay the loan back in 24 monthly installments of $150. How much interest did he pay?

_____

35. Nathan is paid on a piecework basis of $.15 per piece. On Monday he completed 700 pieces; Tuesday, 670; Wednesday, 720; Thursday, 800; and Friday, 760. What was his gross pay for the week?

_____

Reviewing Basic Math Skills

Numbers play an important role in the business world. Numbers, as well as words, are needed to effectively communicate the many activities that take place in the business world daily. In this unit you will have an opportunity to review the basic math skills that you will find necessary for today's business activities.

Reviewing Addition

Addition, the process of computing sets of numbers to find their sum, is an essential skill that people use daily. In business, addition is widely used to solve many problems.

Whole Numbers

In addition, two or more numbers, called **addends**, are combined to find a **sum**. If the numbers in a column total more than 9, carry the first number of the total over to the next column to the left.

```
  1 1 1     carry
  8,679     addends
+ 9,748
 18,427     sum
```

```
  2 0 1 2   carry
  7,357     addends
  9,289
+ 5,318
 21,964     sum
```

To prove the accuracy of your addition, add the same addends working from the bottom of the problem to the top.

	Problem		*Proof*
Adding from top to bottom	1,526 3,244 7,890 +6,110 18,770	Adding from bottom to top	1,526 3,244 7,890 +6,110 18,770

Self-Check

Directions: Find the sum in each problem. Prove each answer in the space provided to the right of each problem. Then compare your answers with those in the back of the book.

1. 54
    +23
    77

2. 42
    +18
    60

3. 945
    + 97
    6042

4. 528
    +393
    921

5. 187
  +625
  _812_ *(handwritten)*

6. 844
  +376
  _1,220_ *(handwritten)*

7. 643
  +287
  _930_ *(handwritten)*

8. 1,399
  +4,710
  _6,109_ *(handwritten)*

9. 9,037
  +1,841
  _10,878_ *(handwritten)*

10. 9,520
  +1,379
  _10,899_ *(handwritten)*

11. 3,924
  +5,063
  _8,987_ *(handwritten)*

12. 2,840
  +1,152
  _3,992_ *(handwritten)*

13. 1,481
  +2,317
  _3,798_ *(handwritten)*

14. 8,495
  +1,417
  _9,912_ *(handwritten)*

15. 5,783
  +6,535
  _12,318_ *(handwritten)*

16. 10,571
  + 6,302
  _16,873_ *(handwritten)*

Decimals

In addition of numbers containing **decimals**, the decimals must be lined up. If necessary, rewrite the numbers in a vertical form. Line up the decimals. Supply zeros as placeholders where needed.

$$1.5 + 0.33 + 94$$

$$
\begin{array}{r}
1.50 \\
0.33 \\
+94.00 \\
\hline
95.83
\end{array}
$$ ← *placeholders*

Self-Check

Directions: Find the sum in each problem. Supply zeros as placeholders where needed. Prove each answer in the space provided to the right of each problem. Then compare your answers with those in the back of the book.

1. 4.3
  +0.295
  _4.595_ *(handwritten)*

2. 934
  1.47
  + 31.3
  _966.77_ *(handwritten)*

3. 1,161.005
  100.1
  +   22.8
  _1,283.905_ *(handwritten)*

4. 0.005
  9.4
  +0.16
  _9.565_ *(handwritten)*

5. 1.1
    0.88
    +0.0011
    _1.9811_

6. 0.0019
    331.006
    + 21.201
    _352.2089_

7. 6.6 + 3.519
    _10.119_

8. 36 + 4.99
    _40.99_

9. 9.3381 + .6616
    _9.9997_

10.   317.83
    +161.16
    _478.99_

11.   821.76
    +178.23
    _999.89_

12.   504.76
    +296.36
    _801.12_

13.   973.573
      81.91
    +   9.20
    _1064.683_

14.   69
      27.198
    +178.26
    _274.458_

15.   0.64
      0.378
    +0.2233
    _1.2413_

16.   135.06
      0.63
    + 42.
    _177.69_

Practice

Directions: Find the sum in each problem. Prove each answer in the space provided to the right of each problem. Then compare your answers with those in the back of the book.

1.   2,642
    +4,135
    _6,777_

2.   7,921
    +2,639
    _10,560_

3.   7,884
    +7,069
    _14,953_

4.   46,234
    +11,325
    _57,559_

5.   17,694
    +15,893
    _33,587_

6.   37,491
    +21,308
    _58,799_

7.   59,641
    +27,840
    _87,481_

8.   9,100
      536
    +2,413
    _12,049_

9. 54.101 + 8.30
    _62.401_

10. .589 + 1.32
    _1.909_

Reviewing Subtraction

Subtraction, the process of deduction, is another important mathematical operation that people use daily. Many common business problems can be solved by using subtraction.

Whole Numbers

Subtraction is the process of taking one number, the **subtrahend**, from another number, the **minuend**, to find the **difference**. When one number in the column is too large to be subtracted from the other number, borrow ten from the next column to the left.

6 16 ← *borrow*	6 12 ← *borrow*
7 6 ← minuend	9 6 7 2 ← minuend
− 4 9 ← subtrahend	− 4 1 3 6 ← subtrahend
2 7 ← difference	5,5 3 6 ← difference

To prove the accuracy of your subtraction, add the subtrahend to the difference. The sum should equal the minuend.

	Problem		*Proof*
	379		163
Subtracting	− 216	Adding	+ 216
	163		379

Self-Check

Directions: Find the difference in each problem. Prove each answer in the space provided to the right of each problem. Then compare your answers with those in the back of the book.

1. 96
    − 84
    12

2. 72
    − 53
    19

3. 57
    − 33
    24

4. 659
    − 217
    442

5. 436
    − 223
    213

6. 5,792
    − 2,481
    3,311

7. 6̶7̶7̶
    − 28
    649

8. 3̶7̶1̶
    − 73
    298

9. 8̶,̶4̶7̶3̶
    − 2,580
    5,893

10. 3̶,̶0̶0̶7̶
    − 1,009
    1,998

Decimals

In subtraction of numbers containing **decimals**, the decimals must be lined up. When necessary, rewrite the numbers in a vertical form. Line up the decimals. Supply zeros as placeholders where needed.

$$882.163 - 32.1$$

$$
\begin{array}{r}
882.163 \\
-\ 32.1 \\
\hline
\end{array}
$$

$$
\begin{array}{r}
882.163 \\
-\ 32.100 \\
\hline
850.063 \\
\end{array}
$$
←

placeholders

Self-Check

Directions: Find the difference in each problem. Supply zeros as placeholders where needed. Prove each answer in the space provided to the right of each problem. Then compare your answers with those in the back of the book.

1. 7.21
    − 0.11
    7.10

2. 4,464.1
    −   22.
    4,442.1

3. 0.9717
    − 0.31
    0.6617

4. 26,654.8
    −   440.1
    26,214.7

5. 3.155
    − 0.101
    3.054

6. 199.99
    − 119.39
    80.60

7. 3,266.616 − 1,171.2
    2,095.416

8. 93.111 − 21.22
    71.891

9. 7.8 − 2.910
    4.910

10. 65.26
    − 14.31
    50.95

Practice

Directions: Find the difference in each problem. Prove each answer in the space provided to the right of each problem. Then compare your answers with those in the back of the book.

1. 6,140
   − 3,157
   ‾‾‾‾‾‾
    2,963

2. 8,005
   − 6,246
   ‾‾‾‾‾‾
    1,759

3. 7,000
   − 5,432
   ‾‾‾‾‾‾
    1,568

4. 9,392
   − 9,286
   ‾‾‾‾‾‾
    106

5. 9,811
   − 769
   ‾‾‾‾‾‾
    9,042

6. 9,148
   − 958
   ‾‾‾‾‾‾
    8,190

7. 10,743
   − 7,842
   ‾‾‾‾‾‾‾
    2,901

8. 16,942
   − 14,523
   ‾‾‾‾‾‾‾
    2,419

9. 22,493
   − 5,967
   ‾‾‾‾‾‾‾
    16,526

10. 23.10
   − 6.59
   ‾‾‾‾‾‾
    16.51

11. 87.56
   − 82.47
   ‾‾‾‾‾‾
    5.09

12. 39.27
   − 18.38
   ‾‾‾‾‾‾
    20.89

Reviewing Multiplication

Multiplication, like addition and subtraction, is a basic mathematical process that is used to determine amounts in many daily transactions.

Whole Numbers

Multiplication is the process of taking one number, the **multiplier**, times another number, the **multiplicand**, to find the **product**.

$$
\begin{array}{r}
342 \text{ multiplicand} \\
\times 211 \text{ multiplier} \\
\hline
342 \rightarrow 1 \times 342 = 342 \\
3\ 420 \rightarrow 10 \times 342 = 3,420 \\
+68\ 400 \rightarrow 200 \times 342 = 68,400 \\
\hline
72,162 \rightarrow 342 \times 211 = 72,162 \leftarrow \text{product}
\end{array}
$$

To prove the accuracy of your multiplication, reverse the multiplier and multiplicand and multiply again.

Problem	Proof
152	343
× 343	× 152
456	686
608	1715
456	343
52,136	52,136

Self-Check

Directions: Find the product in each problem. Prove each answer in the space provided to the right of each problem. Then compare your answers with those in the back of the book.

1. 52
    × 23
    1,196

2. 62
    × 76
    4,712

3. 71
    × 43
    3,053

4. 83
    × 37
    3,071

5. 465
    × 571
    265,515

6. 878
    × 207
    181,746

7. 470
    × 129
    60,630

8. 231
    × 122
    28,182

9. 736
    × 524
    385,664

10. 480
    × 541
    259,680

11. 437
    × 571
    249,527

12. 542
    × 168
    91,056

13. 8,361
    × 3,723
    31,128,003

14. 7,501
    ×   447
    3,352,974

15. 5,327
    ×   312
    1,662,024

16. 3,251
    ×   515
    1,674,265

When you multiply by 10, add a zero to the right of the multiplicand. When you multiply by 100, add two zeros. When you multiply by 1,000, add three zeros, and so on.

$$\begin{array}{c}135\\ \times\ 10\end{array} = 1,350 \qquad \begin{array}{c}265\\ \times 100\end{array} = 26,500 \qquad \begin{array}{c}978\\ \times 1,000\end{array} = 978,000$$

Self-Check

Directions: Find the product in each problem. Prove each answer in the space provided to the right of each problem. Then compare your answers with those in the back of the book.

1. $\begin{array}{r}73\\ \times 10\\ \hline 730\end{array}$

2. $\begin{array}{r}122\\ \times 10\\ \hline 1,220\end{array}$

3. $\begin{array}{r}345\\ \times 10\\ \hline 3,450\end{array}$

4. $\begin{array}{r}918\\ \times 10\\ \hline 9,180\end{array}$

5. $\begin{array}{r}61\\ \times 100\\ \hline 6100\end{array}$

6. $\begin{array}{r}846\\ \times 100\\ \hline 84,600\end{array}$

7. $\begin{array}{r}1,006\\ \times\ 100\\ \hline 100,600\end{array}$

8. $\begin{array}{r}2,076\\ \times\ 100\\ \hline 207,600\end{array}$

9. $\begin{array}{r}631\\ \times 1,000\\ \hline 631,000\end{array}$

10. $\begin{array}{r}186\\ \times 1,000\\ \hline 186,000\end{array}$

11. $\begin{array}{r}7,210\\ \times 1,000\\ \hline 7,210,000\end{array}$

12. $\begin{array}{r}3,080\\ \times 1,000\\ \hline 3,080,000\end{array}$

Decimals

The number of decimal places in the product is equal to the total number of decimal places in the multiplicand and multiplier.

$$\begin{array}{r}6.24\\ \times\ 2.1\\ \hline \end{array} \rightarrow \begin{array}{r}6.24\ \leftarrow 2\text{ decimal places}\\ \times\ 2.1\ \leftarrow 1\text{ decimal place}\\ \hline 13.104\ \leftarrow 3\text{ decimal places in product}\end{array}$$

$$\begin{array}{r}0.06\\ \times 0.03\\ \hline \end{array} \rightarrow \begin{array}{r}0.06\ \leftarrow 2\text{ decimal places}\\ \times 0.03\ \leftarrow 2\text{ decimal places}\\ \hline .0018\ \leftarrow 4\text{ decimal places in product}\end{array}$$

Self-Check

Directions: Find the product in each problem. Prove each answer in the space provided to the right of each problem. Then compare your answers with those in the back of the book.

1. 1.91
    ×0.03
    0.0573

2. 0.621
    × 0.8
    0.4968

3. 0.006
    × 0.09
    0.00054

4. 1.44
    ×0.31
    0.4464

5. 7.698
    × 1.03
    7.92894

6. 32.101
    ×   9.9
    317.7999

7. 1.01
    ×0.99
    0.9999

8. 0.331
    ×0.116
    0.038396

9. 6.715
    × 9.03
    60.63645

10. 1.334
     × .68
     0.90712

Practice

Directions: Find the product in each problem. Prove each answer in the space provided to the right of each problem. Then compare your answers with those in the back of the book.

1. 73
    ×21
    1,533

2. 42
    ×22
    924

3. 212
    ×412
    87,344

4. 7,143
    × 102
    728,586

5. 8,643
    ×   501
    4,330,143

6. 5,327
    × 312
    1,662,024

7. 7,345
    ×   10
    73,450

8. 2,889
    × 100
    288,900

9. 5,486
    ×1,000
    5,486,000

10. 8.36
     × 1.5
     12.54

Reviewing Division

Division is the operation of determining how many times one quantity is contained in another. Division is another important mathematical function that is widely used in business.

Whole Numbers

In division, one number, called the **dividend**, is divided by another number, called the **divisor**, to find the **quotient**.

$$
\begin{array}{c}
\text{divisor} \\
\downarrow \\
27\overline{)4941} \\
\uparrow \\
\text{dividend}
\end{array}
\rightarrow
\begin{array}{r}
1 \\
27\overline{)4941} \\
-27 \\
\hline
22
\end{array}
\rightarrow
\begin{array}{r}
18 \\
27\overline{)4941} \\
-27\downarrow \\
\hline
224 \\
-216 \\
\hline
8
\end{array}
\rightarrow
\begin{array}{r}
183 \leftarrow \text{quotient} \\
27\overline{)4941} \\
-27\downarrow \\
\hline
224 \\
-216\downarrow \\
\hline
81 \\
-81 \\
\hline
\end{array}
$$

To prove the accuracy of your division, multiply the divisor by the quotient.

Problem	*Proof*
$\begin{array}{r} 105 \\ 378\overline{)39{,}690} \\ 378 \\ \hline 1890 \\ 1890 \\ \hline \end{array}$	$\begin{array}{r} 378 \\ \times 105 \\ \hline 1890 \\ 3780 \\ \hline 39{,}690 \end{array}$

The divisor may not divide evenly into the dividend. Therefore, the quotient will have a **remainder** that may be shown as a fraction.

$$
\begin{array}{r}
1 \\
48\overline{)684} \\
-48 \\
\hline
20
\end{array}
\longrightarrow
\begin{array}{r}
14\frac{12}{48} \\
48\overline{)684} \\
-48 \\
\hline
204 \\
-192 \\
\hline
12
\end{array}
\quad = 14\frac{1}{4}
$$

remainder $\longrightarrow$ 12

reduce to lowest terms

Self-Check

Directions: Find the quotient in each problem. Prove each answer in the space provided at the right of each problem. Then compare your answers with those in the back of the book.

1. $6\overline{)321}$ 53.5

2. $8\overline{)705}$ 88.125

3. $19\overline{)1,197}$ 63

4. $15\overline{)211}$ 14.06666667

5. $31\overline{)45,415}$ 1,465

6. $22\overline{)751}$ 34.13636364

7. $311\overline{)7,153}$ 23

8. $55\overline{)6,050}$ 110

9. $42\overline{)8,251}$ 196.452381

10. $92\overline{)7,005}$ 76.14130435

11. $19\overline{)7,239}$ 381

12. $51\overline{)2,563}$ 50.25490196

13. $101\overline{)25,553}$ 253

14. $901\overline{)48,654}$ 54

15. $760\overline{)28,880}$ 38

16. $622\overline{)32,569}$ 52.36173633

Decimals

If there is a decimal in the divisor, move the decimal to the right to make the divisor a whole number. Then move the decimal in the dividend to the right the same number of places. Position the decimal in the quotient directly above the decimal in the dividend. If needed, add zeros to the right of the decimal point in the dividend.

$$6.6\overline{)27.72} \;\rightarrow\; 6\,.\,6.\,)\,27\,.\,7.2 \;\rightarrow\; \begin{array}{r} 4.2 \\ 66\overline{)277.2} \\ -264\;\downarrow \\ \hline 13\;2 \\ -13\;2 \\ \hline \end{array}$$

move one decimal place

When the dividend is an amount of money, place the dollar sign in the quotient and round to the nearest cent.

$$30\overline{)\$36.63}$$

$$\begin{array}{r} \$\ 1.221 = \$1.22 \\ 30\overline{)\$36.630} \\ \underline{30} \\ 66 \\ \underline{60} \\ 63 \\ \underline{60} \\ 30 \\ \underline{30} \end{array}$$

When dividing by 10, 100, or 1,000, count the number of zeros in 10, 100, or 1,000 and move the decimal point in the dividend to the left the same number of places to get the answer.

$$10\overline{)4.8} = 0.48 \qquad 100\overline{)23.01} = .2301 \qquad 1{,}000\overline{)1343.6} = 1.3436$$

Self-Check

Directions: Find the quotient in each problem. Prove each answer in the space provided. Round dollar amounts to the nearest cent. Then compare your answers with those in the back of the book.

1. $2.5\overline{)11.5}$ *4.6*

2. $3.2\overline{)18.272}$ *5.71*

3. $3.15\overline{)53.55}$ *17*

4. $4.08\overline{)26.52}$ *6.5*

5. $3.8\overline{)4.56}$ *1.2*

6. $24.12\overline{)369.036}$ *15.3*

7. $39\overline{)\$282.75}$ *$7.25*

8. $25\overline{)\$29.90}$ *$1.20*

9. $85\overline{)\$134.47}$ *$1.58*

10. $10\overline{)4.9}$ *0.49*

11. $10\overline{)87.88}$ *8.788*

12. $100\overline{)582.5}$ *5.825*

13. 6.3846
100)638.46

14. 9.96458
1,000)9,964.58

15. 0.743256
1,000)743.256

16. 0.086432
1,000)86.432

Practice

Directions: Find the quotient in each problem. Prove each answer in the space provided. Round dollar amounts to the nearest cent. Then compare your answers with those in the back of the book.

1. 20.375
32)652

2. 23
81)1,863

3. $$316.46$
48)$15,190

4. 540
27)14,580

5. $84.08602/51$
93)7,820

6. 34
603)20,502

7. 3.5
3.02)10.57

8. 245
0.136)33.32

9. 620
0.21)130.2

10. 2.35
7.2)16.920

11. 450
0.032)14.400

12. $$36.50$
0.28)$10.22

13. 53.72
10)537.2

14. 36.71
100)3,671.0

15. 57.6221
1,000)57,622.1

16. 211.695
1,000)211,695

Unit 1 Review

Addition

Directions: Find the sum in each problem.

1.   78
     +85
    163

2.   29
     +94
    123

3.   314
     + 37
    351

4.   657
     +349
    1,006

5.   415
     +986
    1,401

6.   9,569
     +3,784
    13,353

7.   2,651
     +1,987
    4,638

8.   6,799
     +2,478
    9,277

9.   4,567
     +9,796
    14,363

10.   5,975
     +8,358
    14,333

11. $3.5 + 7.22 + .59 =$ *11.31*

12. $9.7 + 2.1 + 5.002 =$ *16.802*

13.   2.03
      4.54
     +8.007
    14.577

14. $685.007 + 10.37 + .71 =$ *696.087*

15. $87. + 1.3589 + 274.0075 =$ *362.3664*

16. $401.0041 + 700.39503 + .70001 =$ *742.09914*

17. $54.301 + 723.06 + 8.759 =$ *786.12*

18. $.75 + .7005 + 7.005 =$ *8.4555*

19.   148.810
      221.097
     +173.206
    543.113

20.   18.009
     149.910
     +251.295
    419.214

Directions: Find the difference in each problem.

21. 57
    −49
    8

22. 48
    −19
    29

23. 62
    −39
    23

24. 482
    − 98
    384

25. 531
    − 64
    467

26. 402
    −139
    263

27. 561
    −549
    12

28. 17,003
    − 8,995
    8,008

29. 14,212
    − 9,876
    4,336

30. 11,234
    − 9,549
    1,685

31. 2.3 − 1.9 = *.4*

32. 5.41
    −3.52
    1.89

33. 3.57 − 1.92 = *1.65*

34. 722.31
    − 59.63
    662.68

35. 52.563
    −39.697
    12.866

36. 4.2134 − .9785 =
    3.2349

37. 79.22 − 56.34 =
    22.88

38. 37,433.21
    −   567.02
    36,866.19

39. 21.352
    − 19.86
    1.492

40. 42,111
    − 9,887
    32,224

Multiplication

Directions: Find the product in each problem.

41.   49
    × 37

1,813

42.   92
    × 85

7,820

43. 21 × 15 = *315*

44.   631
    × 592

373,552

45.   500
    × 99

49,500

46.   6,311
    × 295

1,861,745

47. 7,973
    × 220

1,754,060

48. 3,114
   × 6,978

21,729,492

49. 5,607
   × 1,468

8,231,076

50. 7,430
    × 600

4,458,000

51.   2.36
   × 7.21

17.0156

52. .351 × 6.98 =

2.44998

53. 29.578
    × 9.9

292.8222

54. 2.311 × .89 =

2.05679

55.   98.7
   × 4.6

454.02

56. 923.1 × 6.7 =

6,184.77

57.   .457
   × 9.6

4.3872

58. 654.3
   × .52

340.236

59. 0.534
  × 0.291

0.155394

60. 14.923
   × 0.76

11.34148

Division

Directions: Find the quotient in each problem.

61. 7)490   *70*

62. 35)595   *17*

63. 28)3,948   *141*

64. 25)5,960   *238.4*

65. 56)36,792   *657*

66. 63)25,956   *412*

67. 374)34,408   *92*

68. 311)19,904   *64*

69. 114)1,837   *16.1140350 9*

70. 830)24,070   *29*

71. 10).32   *0.032*

72. 78)2,106   *27*

73. 10)9.362   *.9362*

74. 9.08)58.8384   *6.48*

75. 21.1)206.78   *9.8*

76. 100)251.369   *2.51369*

77. 65.9)718.31   *10.9*

78. 1,000)846.925   *.846925*

79. 5.6)52.416   *9.36*

80. 8.34)7.416   *.889208633*

UNIT 2

Reviewing Basic Math Principles

Imagine the problems we would face every day if we couldn't use math. What would we do if we couldn't multiply, add, or subtract? What would happen if we couldn't apply basic math principles to solving problems? The world of business and even our individual world wouldn't come to a halt, but they would work less effectively. Knowing how to use basic math principles makes problem solving a lot easier!

19

Places and Values of Whole Numbers and Decimals

Each digit in a number has a place and a value. The location of a digit within a number is called its place. **Place value** is the value of a digit based on its location within a number group. For example, the place value chart shows the number groups, and within each group, the place value of each digit, depending upon the location of the digit within the group. Look at the place value chart.

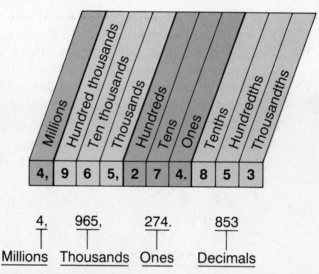

| 4, | 9 | 6 | 5, | 2 | 7 | 4. | 8 | 5 | 3 |

4,	965,	274.	853
Millions	Thousands	Ones	Decimals

Numbers to the left of the decimal point are whole numbers. Numbers to the right of the decimal have values of tenths, hundredths, thousandths of a whole number, and so on. Look at the number 4,965,274.853 from the chart and determine the place value of each digit.

Digit	Place	Value
4	millions	4,000,000 (4 × 1,000,000)
9	hundred thousands	900,000 (9 × 100,000)
6	ten thousands	60,000 (6 × 10,000)
5	thousands	5,000 (5 × 1,000)
2	hundreds	200 (2 × 100)
7	tens	70 (7 × 10)
4	ones	4 (4 × 1)
8	tenths	$\frac{8}{10}$ or 0.8 (8 × .1)
5	hundredths	$\frac{5}{100}$ or 0.05 (5 × 0.01)
3	thousandths	$\frac{3}{1000}$ or 0.003 (3 × 0.001)

Self-Check

Directions: Give the place and value of each underlined digit. Then compare your answers with those in the back of the book.

		Place	*Value*			*Place*	*Value*
1.	**4**83	Hundreds	400	2.	**2**9	tens	20
3.	5.9**8**	Hundredths	0.08	4.	**3**,210	thousands	3,000
5.	63.**4**5	tenths	.4	6.	4.74**7**	thousandths	.007

Writing Numbers Accurately

Writing and copying numbers accurately requires that you understand what numbers mean. Knowing how numbers are written and where they are placed in a group is important for accuracy, too. Numbers can be expressed in either words or figures. For example, when writing a check it is necessary to write the amount in figures, then words. The place value chart can help you write numbers accurately.

EXAMPLE

If someone told you your bank balance was one thousand, three hundred twenty-four dollars and twenty-four cents, would you know how to write this amount in figures? Yes, you would write it as follows:

one thousand, three hundred twenty-four dollars and
twenty-four cents = $1,324.24

When writing numbers, use a comma to separate thousands, millions, and billions in numbers of four or more digits. The rule applies to numbers both in words and in figures. Also use a hyphen in spelled-out compound numbers from twenty-one to ninety-nine. Some examples follow:

five thousand, four hundred fifty-one and six tenths = 5,451.6
ten thousand, two hundred dollars and fifty-nine cents = $10,200.59

Self-Check

Directions: Change the following written amounts to figures. Then compare your answers with those in the back of the book.

1. Ten thousand, nine hundred ninety-one dollars and fifty cents

   $10,991.50

2. Ten thousand, nine hundred ninety-one and five tenths

   10,991.5

3. Four hundred sixty-five thousand, forty-eight dollars and forty-six

   cents $465,048.46

4. One million, six hundred thousand, forty-eight and nine hundredths

   1,600,048.09

Figures to Words

Often it is necessary to convert figures into words. For example, the number $3,414.32 is expressed in words as follows:

$3,414.32 = three thousand, four hundred fourteen dollars and thirty-two cents

Self-Check

Directions: Change the following amounts from figures to words. Then compare your answers with those in the back of the book.

1. 66.19 _Sixty-Six and Nineteen Hundredths_

2. $409.32 _four hundred nine dollars and thirty-two cents_

3. 25,659 _twenty-five thousand, Six hundred and fifty-nine_

4. $1,242.09 _one thousand, two hundred forty-two dollars and nine cents_

5. 592,122 _five hundred ninety-two thousand, one hundred and twenty-two_

6. $24,914,279.75 _twenty-four million, nine hundred fourteen thousand, two hundred seventy-nine dollars and seventy-five cents_

Copying Numbers Accurately

In transferring numbers from one place or form to another, errors can occur. The most common error is reversing the order of numbers. Here are a few examples of common errors:

911,210 becomes 911,120
4,123 becomes 4,132
5 can become 3

Proofreading your work carefully when working with figures can help prevent careless errors.

Self-Check

Directions: Proofread the following sets of numbers. Write X in the space provided if the two amounts are not identical. Then compare your answers with those in the back of the book.

1. _____ 4,281,208   4,281,208

2. ___X___ 611,012   610,012

3. ___X___ 423,116,021   423,161,021

4. _____ 892,204   892,204

5. ___X___ 189,298   198,298

6. ___X___ 6,277,202   6,277,022

Practice

Directions: Follow the directions for each set of problems. Then compare your answers with those in the back of the book.

Give the place and value of each underlined digit.

	Place	Value			Place	Value
1. 3.00432	thousandths	.004		2. 43,128	thousands	3,000

3. <u>6</u>45   hundreds   _600_   4. <u>1</u>,280,000   millions   _1,000,000_

5. <u>7</u>,432   thousands   _7,000_   6. <u>8</u>8   tens   _80_

7. 90.<u>4</u>0   tenths   _.40_   8. 9,780.0<u>9</u>8   thousandths   _.09_

9. <u>7</u>67,897.09   hundred thousands   _700,000_   10. 34.<u>1</u>2   ones   _4_

Change the following numbers to words.

11. $4,903.54 _four thousand, nine hundred three dollars and fifty-four cents_

12. $28,996.32 _twenty-eight thousand, nine hundred ninety-six dollars and thirty-two cents_

13. 102,462 _one hundred two thousand, four hundred and sixty-two_

14. 4,201,197 _four million, two hundred one thousand, one hundred ninety-seven_

15. 641,203,837.093 _six hundred forty-one million, two hundred three thousand, eight hundred thirty-seven and ninety-three thousandths_

Write the following amounts in figures.

16. Five thousandths _.005_

17. One hundred and one tenth _100.1_

18. Six and nine tenths _6.9_

19. Five thousand sixty-five dollars and thirteen cents _$5,065.13_

20. Ten thousand and two tenths _10,000.2_

Rounding Whole Numbers and Decimals

In estimating answers to math problems, one short cut is to round numbers. For instance, Terry knew the distance to the state capital was 196 miles. However, when she wanted to determine how long it would take her to get there if she traveled at a speed of 50 miles per hour, she used 200 miles. Terry rounded the distance to the nearest hundred to make it easier to divide.

Rounding Whole Numbers

Before rounding a number, first determine the desired rounding place—ones, tens, hundreds, and so forth. If the digit to the right of the desired rounding place is 5 or more, add 1 to the digit in the desired rounding place and replace succeeding digits with zeros.

EXAMPLE   Round 159 to the nearest ten.

Step 1   The desired rounding place is the 5. The 5 is   159
          in the tens' place and has a value of 50.

Step 2   The digit to the right of the desired rounding   159
          place, 9, is more than 5.

Step 3   Round up by adding 1 to the number in the   160
          tens' place. Change all digits to the right of
          the desired rounding place to zeros.

If the digit to the right of the rounding place is 4 or less, leave the digit as
it is and change the succeeding digits to zeros.

EXAMPLE   Round 2,148 to the nearest hundred.

Step 1   The digit to the right of the desired rounding   2,148
          place, 4, is less than 5.

Step 2   Round down to the nearest hundred.   2,100
          Replace digits to the right with zeros.

As you can see in the following example, you can round numbers to a
variety of rounding places.

EXAMPLE   Round 5,620,758,781 to the nearest:

Ten: 5,620,758,780
Hundred: 5,620,758,800
Thousand: 5,620,759,000
Ten thousand: 5,620,760,000
Hundred thousand: 5,620,800,000
Million: 5,621,000,000
Ten million: 5,620,000,000
Hundred million: 5,600,000,000
Billion: 6,000,000,000

Self-Check

Directions: Round each number to the indicated place. Then compare
your answers with those in the back of the book.

Round each number to the nearest ten.

1. 308 _310_   2. 15 _20_   3. 1,679 _1,680_

4. 8,611 _8,610_   5. 49 _50_   6. 440 _440_

Round each number to the nearest hundred.

7. 949 _900_   8. 123,456 _123,500_   9. 2,871 _2,900_

10. 13,481 _13,500_ 11. 1,013 _1,000_   12. 261,398 _261,400_

Round each number to the nearest thousand.

13. 1,797 _2,000_   14. 2,501,623 _2,502,000_ 15. 99,647 _100,000_

16. 235,968 _236,000_ 17. 199,555 _200,000_ 18. 9,811 _10,000_

19. 13,481 _13,000_   20. 1,013 _1,000_   21. 261,398 _261,000_

Round each number to the nearest million.

22. 3,567,098 _4,000,000_ 23. 22,789,098 _23,000,000_ 24. 100,100,100 _100,000,000_

25. 9,560,100 _10,000,000_ 26. 6,786,900 _7,000,000_ 27. 1,111,222,000 _1,111,000,000_

Rounding Decimals

In rounding a decimal, use the same procedure as in rounding whole numbers. Decimals are the numbers to the right of the decimal point.

EXAMPLE   Round 23.115 to the nearest tenth.

Step 1 The digit in the tenths' place (the desired   23.1̲15
rounding place) is 1. The digit to the right is
less than 5.

Step 2 Leave the digit in the desired rounding place   23.100
as is. Do not add 1 to the tenths' place.

Step 3 Drop all digits beyond the tenths' place.   23.1

EXAMPLE   Round 6.009 to the nearest hundredth.

Step 1 A zero is in the hundredths' place   6.00̲9
(the desired rounding place).

Step 2 The digit to the right of the hundredths'   6.01
place is more than 5. Change the 0 to 1
and drop the final digit.

Self-Check

Directions: Round each number to the indicated place. Then compare your answers with those in the back of the book.

Round each number to the nearest tenth.

1. 1.77 _1.8_   2. 128.34 _128.3_   3. 62.900 _62.9_

4. .11 _.1_   5. .65 _.7_   6. .19 _.2_

Round each number to the nearest hundredth.

7. 0.245 _.25_   8. 4.12 _4.12_   9. 5.305 _5.31_

10. 26.08 _26.08_   11. 539.345 _539.35_ 12. 23.0003 _23._

Round each number to the nearest thousandth.

13. 93.654 _93.65_ 14. 3,432.45621 _3,432.456_ 15. 10.33325 _10.333_

16. 1.34333 _1.343_ 17. 4.23412 _4.234_   18. 4,324.0004 _4,324._

Rounding Dollar Amounts

The decimal system is also used for our money. Occasionally you may have to round dollar amounts in money problems. The most common procedure is to round a dollar amount to the hundredths' place.

Dollar Amount	Rounded to Hundredths' Place
$ 33.33333	$ 33.33
$ 56.5671	$ 56.57
$ 4.018	$ 4.02
$126.721	$126.72

You may also need to round a dollar amount to the nearest whole dollar, eliminating cents. For instance, $24.99 becomes $25.00, and $24.49 becomes $24.00. When rounding to the nearest dollar, $.50 or more is rounded up, and $.49 or less is rounded down. Look at the following examples:

Dollar Amount	Rounded to Nearest Dollar
$ 56.69	$ 57.00
$ 10.26	$ 10.00
$113.90	$114.00

Self-Check

Directions: Round each dollar amount to the indicated place. Then compare your answers with those in the back of the book.

Round each number to the nearest hundredth.

1. $34.33342 _$34.33_   2. $5,342.54333 _$5,342.54_   3. $5.555 _$5.56_

4. $98.9999 _$99.00_   5. $3.912 _$3.90_   6. $7.345 _$7.35_

7. $78.326 _$78.33_   8. $112.2123 _$112.21_   9. $75.356 _$75.36_

Round each number to the nearest whole dollar amount.

10. $322.21 _$322.00_   11. $6.76 _$7.00_   12. $44,567.06 _$44,567.00_

13. $109.99 _$110.00_   14. $89.88 _$90.00_   15. $678.08 _$678.00_

16. $24.24 _$24.00_   17. $566.34 _$566.00_   18. $7,009.05 _$7,009.00_

Practice

Directions: Round each dollar amount to the indicated place. Then compare your answers with those in the back of the book.

Round each number to the nearest hundredth.

1. $14.891 _$14.89_   2. $5.789 _$5.79_   3. $90.456 _$90.46_

4. $1.234 _$1.23_   5. $35.345 _$35.35_   6. $56.321 _$56.32_

Round each number to the nearest whole dollar amount.

7. $45.59 _$46.00_   8. $35.67 _$36.00_   9. $123.11 _$123.00_

10. $567.01 _$567.00_   11. $89.91 _$90.00_   12. $426.66 _$427.00_

Comparing Whole Numbers and Decimals

In business, many people work with large amounts of similar figures. Accountants and bookkeepers continuously compare numbers. A sales manager must recognize high or low figures at a glance. In this unit you will learn how to compare numbers quickly.

Whole Numbers and Decimals

To compare numbers to see which one is greater, first find out if one of the numbers has more digits to the left of the decimal point. For example, 100 is greater than 99.62 because 100 has three digits to the left of the decimal point and 99.62 has only two. If both numbers have the same number of digits to the left of the decimal point, compare each digit, beginning at the left. One way is to write the value of each digit.

EXAMPLE

Compare 4,932 and 4,943 to see which number is greater.

$$4,932 = \underbrace{4,000}_{\text{same}} + \underbrace{900}_{\text{same}} + 30 + 2$$
$$4,943 = 4,000 + 900 + 40 + 3$$

Forty is greater than 30, so 4,943 is greater than 4,932.

You may not have to write the value of each digit. Just compare each digit starting at the left. Compare 34,751 and 34,729 to see which number is greater.

$$\underbrace{3\ 4\ ,\ 7}_{\text{same}}\ 5\ 1$$
$$3\ 4\ ,\ 7\ 2\ 9$$

Five is greater than 2, so 34,751 is greater than 34,729.

Self-Check

Directions: Compare these numbers and circle the greater one. Then compare your answers with those in the back of the book.

1. 3,456 or ⟨3,465⟩   2. ⟨15,211⟩ or 15,210   3. ⟨543,245⟩ or 543,244

4. 232,510 or ⟨232,521⟩   5. ⟨654⟩ or 645   6. 10,115 or ⟨11,115⟩

7. ⟨3,579⟩ or 3,564   8. ⟨18,431⟩ or 18,411   9. 523,411 or ⟨532,411⟩

Decimals

When comparing decimals, you may want to write zeros to the right of the decimal point so that the numbers being compared have the same number of decimal places. For example, compare .4 and .439 without writing the value of each digit.

.400   Three is greater than zero,
.439   so .439 is greater than .400.

Self-Check

Directions: Compare these numbers and circle the greater one. Then compare your answers with those in the back of the book.

1. 319.59 or (319.62)   2. 29.791 or (29.794)   3. 710.59 or (710.594)

4. (0.05) or 0.005   5. 19.1 or (19.131)   6. (0.764) or 0.763

Practice

Directions: Follow the instructions for each set of problems. Then compare your answers with those in the back of the book.

Compare these numbers and circle the greater one.

1. (43) and 34   2. (876) and 868   3. 3,501 or (3,510)

4. (5,927) or 5,926   5. (3.87) or 3.78   6. (41.6) or 40.9

Write each set of numbers in order from lowest to highest.

7. 1.47, 1.36, 1.5 _1.36, 1.47, 1.5_   8. 5.13, 5.09, 5.10 _5.09, 5.10, 5.13_

9. 6.08, 6.38, 6.81 _6.08, 6.38, 6.81_   10. 4.86, 4.95, 4.81 _4.81, 4.86, 4.95_

Using a Calculator: *Decimals*

The decimal point or period key makes it very easy to do decimal problems with a calculator. For instance, to multiply 125.65 by 17.34, press **1 2 5 . 6 5 × 1 7 . 3 4 =**. Your answer is 2,178.771. Gerry Ainsley wanted to determine how many miles per gallon (MPG) his new car got. Each time he stopped for gas, he filled the tank and noted his mileage and the number of gallons of gas put in. He used his calculator to figure the mileage and MPG. The results of his calculations are as follows:

Ending Mileage		Beginning Mileage		Miles Driven		Gallons		MPG
939.1	−	555.7	=	383.4	÷	12.9	=	29.72
1,364.7	−	939.1	=	425.6	÷	14.5	=	29.35
1,779.4	−	1,364.7	=	414.7	÷	14.5	=	28.6

Gerry then added the MPG for each fill-up and divided the total by 3 to get an average of 29.22. Gerry rounded that number and determined that his car averaged 29 MPG.

Estimating Results

Sometimes giving an educated guess quickly is a requirement when doing a math problem. This process of "mental math" consists of rounding numbers and then projecting a general range of reasonable answers. For example, suppose you earn $5.25 per hour working as a cashier. You can quickly estimate your weekly pay by rounding $5.25 to $5.00 and then multiplying that by 40 work hours. Your weekly pay would be about $200.00. Whether the operation involves addition, subtraction, multiplication, or division, estimating results is a helpful guide to determining the correct answer.

Addition

Round the numbers in the problem to the highest place and calculate the answer mentally. If all the numbers do not have the same number of digits, round the numbers to the highest place of the smallest number. Then, work the actual problem and compare your estimate with the actual answer. For example, to estimate the sum of 203 and 34, do this:

EXAMPLE

Step 1 Round the numbers in the problem to the same place. These numbers are rounded to the nearest ten.

$$\begin{array}{r} 203 \\ +\ 34 \end{array} \rightarrow \begin{array}{r} 200 \\ +\ 30 \end{array}$$

Step 2 Mentally add the rounded numbers to get the estimate.

$$\begin{array}{r} 200 \\ +\ 30 \\ \hline 230 \end{array}$$

Step 3 Compare the estimate to the actual answer.

$$\begin{array}{r} 203 \\ +\ 34 \\ \hline 237 \end{array}$$

Self-Check

Directions: Round numbers to the nearest ten, and estimate the answers. See if your estimates are in the general range of the correct answers. Then compare your answers with those in the back of the book.

		Estimate	Answer			Estimate	Answer
1.	77 +31	110	108	2.	86 +13	100	99
3.	712 +156	900	868	4.	165 + 70	230	235
5.	720 +395	1,100	1,105	6.	483 +315	800	798

Subtraction

When estimating the answer to a subtraction problem, round the numbers as you did for addition, then subtract to get your answer.

EXAMPLE

Step 1 Round the numbers in the problem to the same place. These numbers are rounded to the nearest hundred.

$$
\begin{array}{r} 6{,}731 \\ -\ 763 \end{array} \rightarrow \begin{array}{r} 6{,}700 \\ -\ 800 \end{array}
$$

Step 2 Mentally subtract the rounded numbers to get the estimate.

$$
\begin{array}{r} 6{,}700 \\ -\ 800 \\ \hline 5{,}900 \end{array}
$$

Step 3 Compare the estimate to the actual answer.

$$
\begin{array}{r} 6{,}731 \\ -\ 763 \\ \hline 5{,}968 \end{array}
$$

Self-Check

Directions: Round numbers to the nearest hundred, and estimate the answers. See if your estimates are in the general range of the correct answers. Then compare your answers with those in the back of the book.

		Estimate	Answer			Estimate	Answer
1.	543 −211	300	332	2.	11,892 − 8,922	3,000	2970
3.	691 −421	300	270	4.	34,673 − 3,777	900	896
5.	24,892 − 5,987	18,900	18,905	6.	111,236 − 90,321	20,900	20,915

Multiplication

When estimating the answer to a multiplication problem, round the numbers in the problem as you did for the previous operations. Then mentally multiply to get your answer.

EXAMPLE

Step 1 Round the numbers in the problem to the same place. These numbers are rounded to the nearest ten.

$$
\begin{array}{r} 78 \\ \times 22 \end{array} \rightarrow \begin{array}{r} 80 \\ \times 20 \end{array}
$$

Step 2 Mentally multiply the rounded numbers to get the estimate.

$$
\begin{array}{r} 80 \\ \times\ 20 \\ \hline 1{,}600 \end{array}
$$

Step 3 Compare the estimate to the actual answer.

$$
\begin{array}{r} 78 \\ \times\ 22 \\ \hline 156 \\ 156 \\ \hline 1{,}716 \end{array}
$$

Self-Check

Directions: Round numbers to the nearest ten, and estimate the answers. See if your estimates are in the general range of the correct answers. Then compare your answers with those in the back of the book.

		Estimate	*Answer*			*Estimate*	*Answer*
1.	12 × 66	700	792	2.	33 × 24	600	792
3.	56 × 43	2,400	2,408	4.	75 × 23	1,600	1,725
5.	87 × 46	4,500	4,002	6.	131 × 28	3,900	3,668

Division

In estimating answers in division you may want to use a somewhat different method of rounding numbers. Suppose you want to divide 313 by 16. If you round the numbers to the nearest ten, you would come up with 310 ÷ 20, which is still not easy to divide in your head. There is an easier way of estimating the answer. The easier way is to use what are called *compatible numbers*. These are numbers that are close to the actual numbers in a problem and are easy to divide in your head. Here are two more examples of compatible numbers:

EXAMPLE

Division Problem	*Compatible Numbers*	*Actual Answer*
16)313	20 15)300	19 16)313 16 153 144 9

EXAMPLE

Division Problem	*Compatible Numbers*	*Actual Answer*
19)7,201	400 20)8,000	379 19)7,201 57 150 133 171 171

Self-Check

Directions: Use compatible numbers to estimate the answers. See if your estimates are in the general range of the correct answers. Then compare your answers with those in the back of the book.

	Estimate	*Answer*		*Estimate*	*Answer*
1. $43\overline{)611}$	15	14.2	2. $22\overline{)523}$	25	23.7
3. $19\overline{)99}$	5	5.2	4. $35\overline{)270}$	7.5	7.7
5. $48\overline{)416}$	8	8.6	6. $16\overline{)4,880}$	245	305

Practice

Directions: Round numbers and estimate the answers. Work the problems to see if your estimates are in the general range of the correct answers. Then compare your answers with those in the back of the book.

		Estimate	*Answer*			*Estimate*	*Answer*
1.	3,456 +1,456	5,000	4,912	2.	56,789 +23,898	81,00	80,687
3.	1,234 +2,456	3,700	3,690	4.	5,678 −3,711	2,000	1,967
5.	981 × 26	29,400	25,506	6.	711 × 12	7,100	8,532
7.	4,567 × 34	137,100	155,278	8.	$51\overline{)310}$	6.2	6.0
9.	$18\overline{)1,543}$	77	85.7	10.	$99\overline{)2,999}$	30	30.29

Help with *Dividing Compatible Numbers*

Compatible numbers are numbers that are close to the actual numbers, but are easy to divide. Fred Yates works in the advertising department of the *Pottstown Daily Telegraph*. Pottstown has a population of 43,210. The *Telegraph* has a circulation of 4,897. Fred estimates that 1 out of every 8 people in Pottstown reads the *Telegraph*. How does he arrive at that figure without using a calculator? Fred estimates by using compatible numbers.

Fred reasons that 43,210 is close to 40,000 and 4,897 is close to 5,000. Using compatible numbers, Fred can easily divide them to estimate the answer:

$$40,000 \div 5,000 = 8$$

Fred can also use compatible numbers to estimate the answers to many other problems. See if you can quickly determine the compatible numbers for the following problems:

- An average family has 2.3 people. How many families live in Pottstown? ($40,000 \div 2$)

- Pottstown has 157 businesses. Sixty-three advertise in the *Telegraph*. What percent of the businesses advertise? ($150 \div 50$)

- A $4\frac{1}{4}$-square inch advertisement sells for \$375. What is the cost per square inch? (\$400 \div 4$)

Section 5

Solving Word Problems

Word problems may include more than one operation and are sometimes complicated. The following *four-step method* can be used to solve word problems as well as other kinds of problems.

Step 1	Read	Read the problem through carefully to answer these questions: a. What information is being asked for? b. What information is given in the problem? c. What information is needed to solve the problem?
Step 2	Plan	Determine how you will solve the problem.
Step 3	Work	Carry out the plan. Do any necessary calculations.
Step 4	Answer	Is the answer reasonable? Did you answer the question?

EXAMPLE   A horse won a 1-mile race with a time of 2 minutes. How fast did the
               horse run?

    Step 1   Read. What is the   How fast did the horse run
                  question?   the race?

                  What facts are given?   The race was a mile long and
                                            it was run in 2 minutes.

    Step 2   Plan. What is needed to   "How fast" usually means
                  solve the problem?   "How many miles per hour?"
                                            The horse ran 1 mile in 2
                                            minutes. How many miles
                                          would it run in 1 hour? An
                                          hour has 60 minutes.

    Step 3   Work.   60 minutes divided by 2
                                              minutes.

$$\begin{array}{r} 30 \\ 2\overline{)60} \end{array}$$

    Step 4   Answer.   The horse ran the race at 30
                                          miles per hour.

Some word problems provide more information than is needed to
solve the problem. Others cannot be solved without additional information. Identifying what is wanted, what is given, and what is needed allows
you to organize the information and plan your solution.

Self-Check

Directions: Use the four-step method for solving the following word
problems. Then compare your answers with those in the back of the book.

1. The Brennans are painting the outside of their house. It will take 2
    painters 5 days to paint the entire exterior of the house. Each
    painter works 8 hours a day at $12 per hour. How much will it
    cost for the painters to work?

    $960

2. Cathe Doherty makes a student loan payment for college of $75 a
    month. How much will she pay in 3 years?

    $2,100

3. Dennis Taylor purchased a pair of running shoes for $60, 2 pairs
    of casual pants for $25 each, and a cotton shirt for $23. How much
    did he spend?

    $133

Practice

Directions: Use the four-step method for solving the following word problems. Then compare your answers with those in the back of the book.

1. In 1980 the metropolitan population (city and surrounding suburbs) of Cleveland was 2,063,729; Baltimore, 2,071,016; Newark, 2,057,468. Which had the highest population? Approximately how many more people lived there than in the second-largest city?
   Baltimore   10,000

2. Al Cowles had $250 to spend on clothes. He bought a jacket for $65, a sweater for $27, and a pair of slacks for $18. How much did he have left? *$140*

3. Marconi Wholesalers sold to 4 stores in the metropolitan area. Sales of a special travel clock to the stores were 573, 489, 628, and 526 units. What was the average number of units sold to each store? *554*

4. Olivia Brownley made $18,240 a year as an assistant buyer in a department store. She was paid twice a month. What was her total pay each pay period? *$760*

5. George Lacatelle was doing a sales report for his company on markets in South America. He had total population figures for the following countries: Peru, 18,712,659; Bolivia, 6,089,320; Colombia, 28,575,025; Brazil, 129,660,795; and Paraguay, 3,110,990. Round the population figures for each country to the nearest 100,000.
   Peru, 18,700,000; Bolivia, 6,100,000; Colombia, 28,600,000; Brazil, 129,700,000; Paraguay, 3,100,000

Section 6

Problem-Solving Strategies

When a problem seems difficult or impossible to solve, there are a number of plans or strategies you can use to find the solution. You have already learned about estimating. Some other strategies you may want to try include: restating the problem, working the problem backward, and using a diagram or sketch.

Breaking the Problem into Smaller Steps

A problem that at first appears difficult or complicated often can be broken into smaller steps and become a simple problem to solve. Breaking down a problem into a series of simpler problems or looking at it in another way may lead to the solution.

EXAMPLE   Doug earned $500 in June, $650 in July, and $875 in August. During the summer he had expenses of $100 for travel and $60 for supplies. At the end of the summer, Doug had saved 25% of his net earnings. How much did Doug save?

Strategy: Break the problem down into several smaller problems.

Step 1 Add Doug's earnings.
$$\begin{array}{r} \$\;\;500 \\ 650 \\ +\;\;\;875 \\ \hline \$2{,}025 \end{array}$$

Step 2 Add Doug's expenses.
$$\begin{array}{r} \$100 \\ +\;\;60 \\ \hline \$160 \end{array}$$

Step 3 Calculate net income.
$$\begin{array}{r} \$2{,}025 \\ -\;\;\;160 \\ \hline \$1{,}865 \end{array}$$

Step 4 Determine savings. $\$1{,}865.00 \times .25 = \$466.25$

Self-Check

Directions: Solve the following problems. Break the problems down into smaller steps in order to find the answers. Then compare your answers with those in the back of the book.

1. Clara needs $480 for a winter ski trip to Colorado. She works part-time and is able to save $20 each week from her pay. How many weeks will she need to work in order to save the $480?

   _24_

2. During 1987, a company spent an average of $120 per month on telephone bills. In 1988 the company spent $160. Assuming the company will have a similar increase, how much should the company estimate the phone bill will be for 1989?

   _$200_

3. A company makes a product at a cost of $2.10 for the first 10,000 manufactured and at a cost of $1.90 for the next 10,000. During the month of August, the company sold 15,500 of the product for $3.50 each. How much did the company earn? _$22,800_

Making a Drawing Sometimes using a drawing or sketch helps make the information that has been given clearer and the problem easier to solve.

EXAMPLE   An office is scheduled to be recarpeted soon. The office is 20 feet × 24 feet overall, but has a small space (3 feet × 3 feet) that will not be carpeted. If carpet is selling at $22 a square yard, how much will it cost to carpet this office?

Strategy: Draw a sketch of the room and the uncarpeted space.

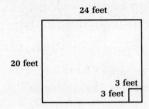

Overall dimension	20 feet × 24 feet = 480 feet
Less uncarpeted area	3 feet × 3 feet = <u> 9</u> feet
	471 feet
Converting to yards	471 feet ÷ 9 (feet in square yards) =
	52.33 square yards
Total cost	52.33 square yards × $22.00 = $1,151.26

Self-Check

Directions: Make a drawing to solve the following problems. Then compare your answers with those in the back of the book.

1. Bob wishes to fertilize his lawn. A bag of lawn fertilizer covers 5,000 square feet. If his property measures 85 feet wide and 125 feet long, how many bags of fertilizer will he need to buy?

   _____3_____

2. Four people ordered three 12-inch pizzas with the agreement that all would share equally. How much would each person receive?

   $\frac{3}{4}$ _____

3. Each sprinkler system waters grass in a circular area with a 5-foot radius. If three sprinklers were placed in a straight line so they would overlap by 6 inches, what length of the lawn would be watered in this design?

   _/4_____

Working Backward

Working backward sometimes becomes an easier and faster way to solve a problem than starting from the beginning, especially if a final result is given. Using the known values, you can turn a problem into a simple arithmetic computation.

EXAMPLE

You need to save $450 by the end of summer for school expenses. Your summer job pays $3.90 per hour, and your "take home" pay is $2.50 per hour. You figure that your other "out-of-pocket" expenses during the summer should be about $300. How many hours must you work to save the $450?

Strategy: First determine the total of what you want to save and what you expect to spend over the summer and then divide by your hourly take-home pay.

$$\$450.00 + \$300.00 ÷ \$2.50 = 300 \text{ hours}$$

Self-Check

Directions: Solve the following problems. Work the problems backward in order to find the answers. Then compare your answers with those in the back of the book.

1. Diane Remandi wants to buy a new watch that costs $65 and an outfit that costs $50. She earns $4.00 an hour and her "take home" pay is $3.50 an hour. How many hours must she work to earn enough money to buy the watch and outfit?

    _33½_

2. Carol paid $18 for a taxi ride from the airport to her home, including a $2 tip. According to the sign posted in the taxi, the fare is based on a charge of $3.00 for the first mile and 50¢ for each additional one-half mile. How many miles was Carol charged for?

    _26_

3. John Rodriguez earns $9.50 an hour and works a forty-hour week. Last week his gross pay was $451.25. For how many hours of work was he paid?

    _45_

Practice

Directions: Use the strategies you have learned to solve these problems. Then compare your answers with those in the back of the book.

1. Tom had three attempts at the pole vault event during a recent track and field meet. His heights for three attempts were $16\frac{1}{2}$ feet; 16 feet, 10 inches; and 17 feet, 1 inch. His average height for his previous five meets is 16 feet, 9 inches. Did he better his past performance? What was the average height for these three recent attempts? _Yes 16 ft. 10 in._

2. You wish to lay bricks on the side of a house that measures 50 feet. Given that each brick is 8 inches long, how many bricks are needed to complete this task? _75_

3. Twelve people shared four 8-inch pumpkin pies. Each person had 1 large piece that was $\frac{1}{4}$ of the pie. How many pieces were left over?

    _4_

4. Suppose that 10 bikes could fit into the space required for 2 cars. How many bicycles could be parked in a lot designed for 45 cars?

    _225_

Unit 2 Review

Developing Your Skills

Directions: Follow the instructions for each set of problems.

Give the place and value of each underlined digit.

	Place	Value
1. 9̲00	hundreds	900
2. 4̲,891	thousands	4,000

Change the following written amounts to figures.

3.   Seventy-six thousand, six hundred forty-eight 76,648

4.   One million, twenty-one thousand, eight 1,021,008

Change the following amounts from figures to words.

5. 30,080,111 thirty million, eighty thousand, one hundred eleven

6. $58,512.39 fifty-eight thousand, five hundred twelve dollars and thirty-nine cents

Proofread the following sets of numbers. Write X in the space provided if the two numbers are not identical.

7. X̲   24,016   24,061

8. _____   276,767   276,767

Round each number to the underlined place.

9. 5̲43 540   10. 5,6̲78 5,680

11. 96̲5 970   12. 6̲78 700

Compare these numbers and circle the greatest one.

13. 5,112   (5,122)   14. 3,989   (3,999)

15. 10,345   (10,435)   16. 1,345,333   (1,354,333)

Round numbers and estimate the answers. Work the problems to see if your estimates are in the general range of the correct answers.

		Estimate	Answer			Estimate	Answer
17.	5,988	6,000	8,333	18.	3,789	300	290
	+2,345				−3,499		

		Estimate	Answer			Estimate	Answer

19.   212   *12,600*   *13,356*   20. 64)6,784   *100*   *106*

        × 63

Solving Word Problems

Directions: In solving a mathematics word problem, ask: (1) What information is being asked for? (2) What information is given in the problem? (3) What information is needed to solve the problem? Use the problem-solving steps and strategies that you have learned.

21. Tony Renaldo works in a regional produce warehouse taking orders over the phone for cartons of eggs. Each carton contains 24 dozen eggs. Tony received the following orders for cartons each day last week: 23, 46, 51, 18, 79. _____

    a. What is the total number of cartons ordered last week? *217*

    b. How many dozen eggs were ordered? *5,208*

    c. What was the total number of eggs ordered? *2,604*

22. Susan Lorenzo has a job interview at 10:30 in the morning. She wants to arrive there 10 minutes early. Susan needs 20 minutes for travel time, and she wants to allow an hour and a half to dress, read the paper, and have breakfast. What time should Susan get up? *8:30*

23. Amy Randolf wants to put down a new floor. The area of the floor is 50 feet by 25 feet. The flooring costs $9 a square yard. How much will it cost Amy to put down the new floor? *$138.89*

24. On a shopping trip, Brian Gomez first spent half his money at the music store, then spent $20 at a shoe store. Next, he spent $5.50 for lunch. After lunch, he spent $64 at a sporting goods shop. He then spent his final $15 on groceries. How much money did Brian have at the beginning of his shopping trip? *$209.*

25. The Smiths plan to carpet 3 rooms with carpeting costing $12.95 a square yard. The rooms measure 13 feet by 11 feet, 10 feet by 12 feet, and 15 feet by 18 feet. How much will the carpeting cost?

     $766.90

26. Derek McLaughlin had saved $750 to furnish his apartment. He checked several furniture store advertisements and came up with the following items: two framed pictures, $61.90; one chair, $157.50; kitchen set, $214.89; two end tables , $139.98; two lamps, $86.35; and a rug, $58.99.

    a. Did Derek have any money left? *Yes*

    b. If so, how much? *29.39*

UNIT 3

Working with Fractions

The world of business uses fractions for many purposes: for sales discounts, for some kinds of material specifications such as lumber, in weights and measurements, and occasionally in expressions of distance. We also use fractions in expressing time: half past the hour, a quarter after, and so on. Since fractions are used so commonly, you should know not only how to work with them but also how to convert them to other numeric forms.

Understanding Key Terms

Fraction means *part of* — actually, part of a whole. In mathematics, a **fraction** is a part of a whole number. If a pound of margarine is divided into four quarters, each quarter represents one fourth of the whole and is expressed as:

$$\frac{1}{4} \quad \begin{matrix} \longleftarrow \text{ numerator} \\ \longleftarrow \text{ denominator} \end{matrix}$$

The top number in a fraction is called the **numerator** and the bottom number, the **denominator**. A fraction such as the preceding example can be read in three ways: "one fourth," "one divided by four," or "one over four." Now let's look at some different kinds of fractions.

Proper Fraction

A **proper fraction** is one in which the value of the numerator is smaller than the value of the denominator. A proper fraction always has a value of less than one. The following examples are proper fractions:

$$\frac{3}{4} \quad \frac{15}{19} \quad \frac{7}{8} \quad \frac{99}{100}$$

Improper Fraction

An **improper fraction** is one in which the value of the numerator is larger than the value of the denominator. An improper fraction always has a value greater than one. The following examples are improper fractions:

$$\frac{14}{9} \quad \frac{7}{5} \quad \frac{18}{13} \quad \frac{100}{99} \quad \frac{5}{4}$$

Mixed Fraction

A **mixed fraction** is one with two parts: a whole number and a fraction. The following examples are mixed fractions:

$$4\frac{1}{3} \quad 2\frac{3}{5} \quad 18\frac{7}{11} \quad 1\frac{1}{100} \quad 23\frac{3}{4}$$

Complex Fraction

A **complex fraction** is one in which either the numerator or the denominator or both are fractions. The following examples are complex fractions:

$$\frac{2\frac{1}{3}}{4} \quad \frac{1}{1\frac{2}{3}} \quad \frac{\frac{1}{3}}{4\frac{1}{8}} \quad \frac{10\frac{11}{12}}{3\frac{4}{5}} \quad \frac{1\frac{1}{2}}{100\frac{6}{7}}$$

Self-Check

Directions: Identify each fraction as proper, improper, mixed, or complex. Then compare your answers with those in the back of the book.

1. $\frac{2}{3}$ _Proper_

2. $\frac{11}{9}$ _improper_

3. $\dfrac{\frac{1}{3}}{2\frac{7}{9}}$ _complex_

4. $\dfrac{7}{8}$ _proper_

5. $3\dfrac{3}{8}$ _mixed_

6. $8\dfrac{11}{12}$ _mixed_

7. $\dfrac{4}{3}$ _improper_

8. $55\dfrac{1}{8}$ _mixed_

Practice

Directions: Identify each fraction as proper, improper, mixed, or complex. Then compare your answers with those in the back of the book.

1. $\dfrac{2\frac{4}{9}}{\frac{3}{4}}$ _complex_

2. $\dfrac{16}{11}$ _improper_

3. $\dfrac{21}{23}$ _proper_

4. $\dfrac{5\frac{3}{9}}{\frac{3}{4}}$ _complex_

5. $\dfrac{2}{1}$ _improper_

6. $\dfrac{12}{6}$ _improper_

7. $\dfrac{\frac{2}{3}}{8}$ _complex_

Section 2

Expressing Fractions in Different Terms

When you multiply the numerator and denominator of a fraction by a number greater than one, you are changing the fraction to **higher terms**. When you divide the numerator and denominator of a fraction by the same whole number greater than one, you are reducing the fraction, or changing it to **lower terms**. You are not changing the value of the fraction. You are simply changing the original fraction to an equivalent fraction.

Equivalent Fractions Fractions that are equal in value, even though the numerators and denominators are not the same, are called **equivalent fractions**. The following fractions are all equivalent fractions:

$$\frac{2}{4} \qquad \frac{4}{8} \qquad \frac{64}{128} \qquad \frac{200}{400}$$

To write equivalent fractions, multiply or divide the numerator and denominator by the same number. Let's look at the following example:

$$\frac{2}{4} \times \frac{2}{2} = \frac{4}{8} \text{ or } \frac{4}{8} \div \frac{2}{2} = \frac{2}{4}$$

Self-Check

Directions: Complete these sets of fractions so that they are equivalent fractions. Then compare your answers with those in the back of the book.

1. $\frac{1}{3} = \frac{?}{18}$ 6

2. $\frac{4}{9} = \frac{16}{?}$ 36

3. $\frac{15}{35} = \frac{3}{?}$ 7

4. $\frac{45}{60} = \frac{3}{?}$ 4

5. $\frac{3}{4} = \frac{21}{?}$ 28

6. $\frac{24}{144} = \frac{1}{?}$ 6

Reducing Fractions to Lowest Terms A fraction is reduced to its **lowest terms** when no number except 1 can be divided exactly into both its numerator and denominator. In other words, both the numerator and denominator must be divided equally by the same number. This can be done using a two-step method. For example, to reduce the fraction $\frac{48}{72}$ to its **lowest terms**, do this:

Step 1 Find a number that can be divided into the numerator and denominator equally. $\frac{48}{72} \div \frac{4}{4} = \frac{12}{18}$

Step 2 Repeat step 1 until the only number that will divide evenly into the numerator and denominator is 1, lowest terms. $\frac{12}{18} \div \frac{2}{2} = \frac{6}{9}$

$\frac{6}{9} \div \frac{3}{3} = \frac{2}{3}$

Self-Check

Directions: Reduce the following fractions to lowest terms. Then compare your answers with those in the back of the book.

1. $\frac{2}{6}$ $\frac{1}{3}$

2. $\frac{10}{30}$ $\frac{1}{3}$

3. $\frac{7}{21}$ $\frac{1}{7}$

4. $\frac{5}{30}$ $\frac{1}{6}$

5. $\frac{6}{15}$ $\frac{2}{5}$

6. $\frac{10}{25}$ $\frac{2}{5}$

Reducing Improper Fractions An improper fraction can also be "reduced" to a whole number if the numerator and denominator can be divided evenly or to a mixed fraction if there is a remainder. Let's reduce $\frac{13}{7}$.

Step 1 Divide the numerator by the denominator. $\frac{13}{7} \rightarrow 13 \div 7 \rightarrow 7\overline{)13}$

Step 2 The quotient is the whole number. The remainder becomes the "new" numerator and the divisor remains the denominator. The result is $1\frac{6}{7}$.

$$\begin{array}{r} 1 \\ 7\overline{)13} \\ \underline{7} \\ 6 \end{array} = 1\frac{6}{7}$$

Self-Check

Directions: Reduce the following improper fractions. Then compare your answers with those in the back of the book.

1. $\frac{15}{3}$ 5   2. $\frac{300}{5}$ 60   3. $\frac{20}{4}$ 5

4. $\frac{66}{33}$ 2   5. $\frac{75}{15}$ 5   6. $\frac{180}{9}$ 20

7. $\frac{224}{32}$ 7   8. $\frac{17}{5}$ $3\frac{2}{5}$ 9. $\frac{31}{13}$ $2\frac{5}{13}$

Converting Mixed Fractions to Improper Fractions

Occasionally it is necessary to change a mixed fraction to an improper fraction. For instance, you might want to know how many "quarters" there are in $16\frac{3}{4}$.

Step 1   Multiply the denominator by the whole number: $4 \times 16 = 64$. Then add the numerator to get the product: $64 + 3 = 67$. $16\frac{3}{4}$

Step 2   Put the product 67 over the "old" denominator, and the improper fraction is $\frac{67}{4}$. $\frac{67}{4}$

Self-Check

Directions: Convert the following mixed fractions to improper fractions. Then compare your answers with those in the back of the book.

1. $9\frac{8}{11}$ $\frac{107}{11}$ 2. $12\frac{1}{3}$ $\frac{37}{3}$ 3. $16\frac{4}{27}$ $\frac{436}{27}$

4. $14\frac{3}{6}$ $\frac{87}{6}$ 5. $21\frac{1}{4}$ $\frac{85}{4}$ 6. $42\frac{3}{7}$ $\frac{297}{7}$

7. $7\frac{13}{15}$ $\frac{118}{15}$ 8. $4\frac{21}{22}$ $\frac{109}{22}$ 9. $3\frac{1}{11}$ $\frac{34}{11}$

Practice

Directions: Reduce these fractions to their lowest terms. When necessary, use the step-by-step process. If the fraction cannot be reduced, circle it. Then compare your answers with those in the back of the book.

1. $\frac{7}{28}$ $\frac{1}{4}$ 2. $\frac{15}{25}$ $\frac{3}{5}$ 3. $\frac{8}{12}$ $\frac{2}{3}$

4. $\frac{84}{140}$ $\frac{3}{5}$ 5. $\frac{17}{11}$ $1\frac{6}{11}$ 6. $\frac{56}{7}$ 8

7. $\frac{24}{34}$ $\frac{12}{17}$ 8. $\frac{30}{36}$ $\frac{5}{6}$ 9. $\frac{261}{8}$ $32\frac{5}{8}$

10. $\frac{280}{298}$ $\frac{140}{149}$ 11. $\frac{59}{5}$ $11\frac{4}{5}$ 12. $\frac{72}{9}$ 8

13. $\frac{45}{81}$ $\frac{5}{9}$ 14. $\frac{10}{45}$ $\frac{2}{9}$ 15. $\frac{155}{75}$ $2\frac{1}{15}$

16. $\frac{213}{12}$ $17\frac{3}{4}$ 17. $\frac{42}{16}$ $2\frac{5}{8}$ 18. $\frac{400}{50}$ 8

19. $\frac{220}{18}$ $12\frac{2}{9}$ 20. $\frac{25}{500}$ $\frac{1}{20}$ 21. $\frac{12}{36}$ $\frac{1}{3}$

Working Fraction Problems

Fractions can be added, subtracted, multiplied, and divided. Often they must be changed to equivalent fractions of higher or lower terms before these operations can be performed.

Adding Fractions

To add fractions with the same denominators, add all of the numerators and put that sum over the common denominator. Reduce the fraction to its lowest terms. For example:

$$\frac{1}{8} + \frac{3}{8} + \frac{2}{8} = \frac{6}{8} = \frac{3}{4} \qquad \frac{23}{9} + \frac{13}{9} = \frac{36}{9} = 4$$

Self-Check

Directions: Solve these addition problems for various kinds of fractions. Then compare your answers with those in the back of the book.

1. $\frac{1}{9}$
    $+\frac{3}{9}$
    $\frac{4}{9}$

2. $\frac{1}{8}$
    $+\frac{5}{8}$
    $\frac{3}{4}$

3. $\frac{7}{11}$
    $+\frac{2}{11}$
    $\frac{9}{11}$

4. $\frac{4}{15}$
    $+\frac{6}{15}$
    $\frac{2}{3}$

5. $\frac{5}{7}$
    $+\frac{4}{7}$
    $1\frac{2}{7}$

6. $\frac{3}{4}$
    $+3\frac{3}{4}$
    $4\frac{1}{2}$

7. $2\frac{3}{10}$
    $+7\frac{9}{10}$
    $10\frac{5}{}$

8. $\frac{5}{6}$
    $\frac{1}{6}$
    $+8\frac{1}{6}$
    $9\frac{1}{6}$

9. $3\frac{9}{25}$
    $+1\frac{12}{25}$
    $4\frac{21}{25}$

10. $\frac{7}{13}$
    $7\frac{2}{13}$
    $+1$
    $8\frac{9}{15}$

Adding Fractions with Different Denominators

To add fractions with different denominators, find the **lowest common denominator (LCD)**. Thus, to add $\frac{1}{3} + \frac{1}{5}$, do the following:

Step 1 Convert $\frac{1}{3}$ and $\frac{1}{5}$ to fractions with a common denominator. In this example, multiply the two denominators to determine a common denominator.

$\frac{1}{3} = \frac{?}{15}$
$+\frac{1}{5} = \frac{?}{15}$

Step 2 Convert to equivalent fractions. Raise the top fraction: 3 into 15 = 5, 5 × 1 = 5. Raise the bottom fraction: 5 into 15 = 3, 3 × 1 = 3.

$\frac{1}{3} = \frac{5}{15}$
$+\frac{1}{5} = \frac{3}{15}$

Step 3 Add the fractions to get your answer. Reduce to lowest terms when necessary.

$\frac{5}{15}$
$+\frac{3}{15}$
$\frac{8}{15}$

Self-Check

Directions: Solve these addition problems for various kinds of fractions. Then compare your answers with those in the back of the book.

1. $\frac{9}{13}$ $+\frac{1}{4}$ $\frac{49}{52}$

2. $\frac{4}{5}$ $+\frac{3}{5}$ $1\frac{2}{5}$

3. $2\frac{5}{12}$ $+7\frac{7}{12}$ 10

4. $\frac{23}{8}$ $+\frac{7}{3}$ $5\frac{5}{24}$

5. $\frac{15}{16}$ $+\frac{9}{12}$ $1\frac{11}{16}$

Subtracting Fractions

To subtract fractions with the same denominators, write the difference of the numerators over the common denominator. Reduce to lowest terms. Let's try it.

Step 1 Denominators are the same. Subtract numerators.

$\frac{7}{12}$ $-\frac{3}{12}$ $\frac{4}{12}$

Step 2 Reduce to lowest terms. $\frac{4}{12} = \frac{1}{3}$

Self-Check

Directions: Subtract the following fractions. Then compare your answers with those in the back of the book.

1. $\frac{7}{8}$ $-\frac{3}{8}$ $\frac{1}{2}$

2. $\frac{3}{5}$ $-\frac{1}{5}$ $\frac{2}{5}$

3. $\frac{8}{11}$ $-\frac{5}{11}$ $\frac{3}{11}$

4. $\frac{5}{16}$ $-\frac{1}{16}$ $\frac{1}{4}$

5. $\frac{7}{9}$ $-\frac{4}{9}$ $\frac{1}{3}$

6. $3\frac{8}{9}$ $-\frac{4}{9}$ $3\frac{4}{9}$

7. $2\frac{6}{7}$ $-\frac{1}{7}$ $2\frac{5}{7}$

8. $6\frac{5}{6}$ $-3\frac{1}{6}$ $3\frac{2}{3}$

9. $12\frac{2}{5}$ $-7\frac{1}{5}$ $4\frac{2}{5}$

10. $9\frac{8}{9}$ $-8\frac{5}{9}$ $1\frac{1}{3}$

Subtracting Fractions with Different Denominators

To find the difference between fractions with different denominators, first find the lowest common denominator. Let's look at an example.

Step 1 Find the LCD. $\frac{14}{15} = \frac{?}{15}$ $-\frac{3}{5} = \frac{?}{15}$

Step 2 Raise fraction to higher terms, then subtract. $\frac{14}{15}$ $-\frac{9}{15}$ $\frac{5}{15}$

Step 3 Reduce to lowest terms. $\frac{5}{15} = \frac{1}{3}$

Self-Check

Directions: Subtract the following fractions. Find the LCD. Then compare your answers with those in the back of the book.

1. $\frac{13}{15}$
   $-\frac{1}{3}$
   $\frac{8}{15}$

2. $\frac{5}{7}$
   $-\frac{1}{2}$
   $\frac{3}{14}$

3. $\frac{1}{2}$
   $-\frac{2}{5}$
   $\frac{1}{10}$

4. $40\frac{5}{6}$
   $-16\frac{1}{10}$
   $24\frac{11}{15}$

5. $1\frac{17}{20}$
   $-1\frac{1}{5}$
   $\frac{13}{20}$

Subtracting Mixed Fractions

Subtracting mixed fractions is similar to adding mixed fractions. Let's do the following problem: $7\frac{2}{3} - 3\frac{1}{6}$.

Step 1 First, change the denominators to a common denominator (if necessary).

$$7\frac{2}{3} = 7\frac{4}{6}$$
$$-3\frac{1}{6} = 3\frac{1}{6}$$

Step 2 Subtract the numerator of the mixed fractions and then the whole numbers.

$$7\frac{2}{3} = 7\frac{4}{6}$$
$$-3\frac{1}{6} = 3\frac{1}{6}$$
$$4\frac{3}{6}$$

Step 3 Write the answer as a mixed number in its lowest terms.

$$4\frac{3}{6} = 4\frac{1}{2}$$

Self-Check

Directions: Subtract the following mixed fractions. Then compare your answers with those in the back of the book.

1. $15\frac{3}{8}$
   $-4\frac{1}{8}$
   $11\frac{1}{4}$

2. $8\frac{7}{9}$
   -7
   $1\frac{7}{9}$

3. $11\frac{2}{3}$
   $-7\frac{9}{15}$
   $4\frac{1}{15}$

4. $9\frac{5}{11}$
   $-8\frac{1}{3}$
   $1\frac{4}{33}$

5. $15\frac{3}{8}$
   $-4\frac{1}{8}$
   $11\frac{1}{4}$

Subtracting Mixed Fractions when Borrowing

Subtracting mixed fractions when borrowing is involved requires more steps. Subtract the fractions first, then subtract the whole numbers. Let's do the following problem: $9\frac{1}{3} - 4\frac{3}{5}$.

Step 1 Find the LCD. Convert to equivalent fractions in higher terms.

$$9\frac{1}{3} = 9\frac{5}{15}$$
$$-4\frac{3}{5} = 4\frac{9}{15}$$

Step 2 The denominators are the same, but the numerator in the minuend $\left(\frac{5}{15}\right)$ is smaller than the numerator in the subtrahend $\left(\frac{9}{15}\right)$. To easily subtract, borrow 1 $\left(\frac{15}{15}\right)$ from the whole number, and 9 becomes 8.

$$\overset{8}{9}\frac{5}{15}$$
$$-4\frac{9}{15}$$

Step 3 Add $\frac{15}{15}$ and $\frac{5}{15}$ to raise the fraction to higher, more manageable terms. The bottom fraction remains the same.

$$\frac{15}{15} + 8\frac{5}{15} = 8\frac{20}{15}$$
$$-4\frac{9}{15} \qquad = 4\frac{9}{15}$$

Step 4 Subtract the fractions and then the whole numbers. Write the answer as a mixed fraction in its lowest terms.

$$8\frac{20}{15}$$
$$-4\frac{9}{15}$$
$$\overline{4\frac{11}{15}}$$

Self-Check

Directions: Subtract the following fractions. Then compare your answers with those in the back of the book.

1. $12\frac{2}{9}$
   $-5\frac{5}{9}$
   $6\frac{2}{3}$

2. $4\frac{1}{6}$
   $-1\frac{5}{12}$
   $2\frac{3}{4}$

3. 18
   $-14\frac{8}{9}$
   $3\frac{1}{9}$

4. $5\frac{3}{11}$
   $-4\frac{7}{11}$
   $\frac{7}{11}$

5. 9
   $-6\frac{7}{9}$
   $2\frac{2}{9}$

6. 10
   $-7\frac{4}{5}$
   $2\frac{1}{5}$

7. $79\frac{5}{48}$
   $-61\frac{7}{16}$
   $17\frac{2}{3}$

8. $17\frac{7}{12}$
   $-12\frac{3}{4}$
   $4\frac{5}{6}$

9. $19\frac{3}{12}$
   $-15\frac{15}{36}$
   $3\frac{2}{3}$

10. $17\frac{17}{54}$
    $-14\frac{48}{54}$
    $2\frac{23}{54}$

Multiplying Fractions

To find the product of two or more fractions, multiply the numerators; then multiply the denominators as shown in the following example.

$$\frac{1}{3} \times \frac{5}{8} = \frac{1 \times 5}{3 \times 8} = \frac{5}{24}$$

Sometimes it is possible to reduce the fraction before multiplying the numerator and denominator. This is called **canceling**, which means to divide by a common factor. Both the numerator and denominator must be divided. Let's multiply $\frac{5}{8} \times \frac{4}{5}$.

Step 1 Cancel out, 5 into 5 = 1; 4 into 8 = 2.

$$\frac{\overset{1}{\cancel{5}}}{\underset{2}{\cancel{8}}} \times \frac{\overset{1}{\cancel{4}}}{\underset{1}{\cancel{5}}}$$

Step 2 Multiply to get your answer.

$$\frac{1}{2} \times \frac{1}{1} = \frac{1}{2}$$

Self-Check

Directions: Multiply the following fractions. Cancel where necessary. Then compare your answers with those in the back of the book.

1. $\frac{3}{7} \times \frac{4}{9}$ $\frac{4}{21}$

2. $\frac{5}{8} \times \frac{1}{3}$ $\frac{5}{24}$

3. $\frac{11}{15} \times \frac{5}{9}$ $\frac{11}{27}$

4. $\frac{7}{9} \times \frac{1}{7}$ $\frac{1}{9}$

5. $\frac{2}{5} \times \frac{11}{18}$ $\frac{11}{45}$

6. $\frac{6}{7} \times \frac{1}{2}$ $\frac{3}{7}$

7. $\frac{1}{6} \times \frac{2}{5}$ $\frac{1}{15}$

8. $\frac{2}{3} \times \frac{7}{8}$ $\frac{21}{12}$

9. $\frac{5}{6} \times \frac{8}{9}$ $\frac{20}{27}$

Working Fraction Problems **49**

Multiplying Mixed Fractions

To multiply mixed fractions, first write the mixed fractions as improper fractions. Then multiply the fractions. Let's multiply $3\frac{3}{4} \times 1\frac{3}{5}$.

Step 1 Convert the mixed fractions to improper fractions.

$$\frac{15}{4} \times \frac{8}{5}$$

Step 2 Cancel out to reduce the fractions.

$$\frac{\overset{3}{\cancel{15}}}{\underset{1}{\cancel{4}}} \times \frac{\overset{2}{\cancel{8}}}{\underset{1}{\cancel{5}}}$$

Step 3 Multiply the fractions and reduce to lowest terms to get your answer.

$$\frac{3}{1} \times \frac{2}{1} = \frac{6}{1} = 6$$

Self-Check

Directions: Multiply the following mixed fractions. Then compare your answers with those in the back of the book.

1. $2\frac{7}{9} \times \frac{3}{28}$
2. $3 \times 4\frac{5}{9}$
3. $1\frac{7}{8} \times 2\frac{3}{5}$
4. $9\frac{1}{3} \times 2\frac{1}{4}$
5. $7 \times 6\frac{1}{4}$
6. $4\frac{2}{3} \times \frac{5}{9}$
7. $2\frac{7}{2} \times 5\frac{1}{9}$
8. $\frac{3}{16} \times 3\frac{5}{3}$
9. $3\frac{9}{4} \times \frac{3}{24}$

Dividing Fractions

To divide fractions, invert (turn upside down) the divisor (the fraction following the division sign) and multiply the numerators and denominators. Reduce to the lowest terms. Let's do the following problem: $\frac{3}{4} \div \frac{7}{8}$.

Step 1 Invert the divisor (the fraction following the division sign).

$$\frac{3}{4} \div \frac{7}{8} = \frac{3}{4} \div \frac{8}{7}$$

Step 2 Change the division sign to a multiplication sign.

$$\frac{3}{4} \times \frac{8}{7}$$

Step 3 Cancel out to reduce the fractions.

$$\frac{3}{\underset{1}{\cancel{4}}} \times \frac{\overset{2}{\cancel{8}}}{7}$$

Step 4 Multiply the reduced fractions and reduce to lowest terms.

$$\frac{3}{1} \times \frac{2}{7} = \frac{6}{7}$$

Self-Check

Directions: Solve these division problems for various kinds of fractions. Then compare your answers with those in the back of the book.

1. $\frac{5}{8} \div \frac{1}{3}$
2. $\frac{13}{15} \div \frac{2}{3}$
3. $\frac{6}{7} \div \frac{2}{5}$
4. $\frac{11}{15} \div \frac{3}{7}$
5. $\frac{6}{7} \div \frac{3}{8}$
6. $\frac{13}{17} \div \frac{1}{3}$
7. $\frac{6}{8} \div \frac{4}{10}$
8. $\frac{5}{9} \div \frac{12}{20}$
9. $\frac{15}{25} \div \frac{3}{5}$

Dividing Mixed Fractions

To divide mixed fractions, first convert them to improper fractions. Inve. the divisor (the fraction following the division sign). Multiply the numerators and denominators and reduce the result to the lowest terms. Let's do the following problem: $6\frac{7}{8} \div \frac{5}{8}$.

Step 1 Change the mixed fraction to an improper fraction.

$$6\frac{7}{8} \div \frac{5}{8} =$$
$$\frac{55}{8} \div \frac{5}{8}$$

Step 2 Invert the divisor. Change the division sign to a multiplication sign.

$$\frac{55}{8} \times \frac{8}{5}$$

Step 3 Cancel out where necessary.

$$\overset{11}{\underset{1}{\cancel{\frac{55}{8}}}} \times \overset{1}{\underset{1}{\cancel{\frac{8}{5}}}}$$

Step 4 Multiply the reduced fractions and reduce the result to lowest terms.

$$\frac{11}{1} \times \frac{1}{1} = \frac{11}{1} = 11$$

Self-Check

Directions: Solve these division problems. Then compare your answers with those in the back of the book.

1. $9\frac{1}{3} \div 2\frac{6}{7}$
2. $15\frac{1}{8} \div 3\frac{2}{3}$
3. $7\frac{1}{2} \div 2\frac{1}{3}$
4. $16\frac{3}{4} \div 2\frac{1}{2}$
5. $14\frac{3}{5} \div 2\frac{1}{2}$
6. $2\frac{3}{9} \div 5\frac{6}{15}$

Practice

Directions: Solve these problems for various kinds of fractions. Then compare your answers with those in the back of the book.

1. $\frac{5}{6} \div \frac{2}{3}$
2. $9\frac{2}{4} \div 1\frac{1}{4}$
3. $\frac{8}{17} \div \frac{24}{51}$
4. $8\frac{1}{8} \div \frac{5}{16}$
5. $4\frac{7}{8} \div 13$
6. $17\frac{4}{5} - 12\frac{8}{15}$
7. $42\frac{6}{37} - 40\frac{8}{37}$
8. $15\frac{3}{5} + 3\frac{4}{9}$
9. $\frac{3}{4} \times \frac{5}{6}$

Section 4

Converting Fractions to Decimals

The process of converting fractions to decimals involves dividing the numerator of the fraction by its denominator. Let's convert $\frac{3}{16}$.

Step 1 The denominator, 16, becomes the divisor and the numerator, 3, becomes the dividend and is divided by 16.

$$\frac{3}{16} = 16\overline{)3}$$

Step 2 Add the decimal point and zeros as needed, then divide.

$$16\overline{)3.0000}^{.1875}$$

Self-Check

Directions: Convert the following fractions to decimals. Then compare your answers with those in the back of the book.

1. $\frac{6}{25}$ 2. $\frac{4}{5}$ 3. $\frac{7}{8}$

4. $\frac{45}{50}$ 5. $\frac{8}{20}$ 6. $\frac{2}{5}$

Repeating Decimal When a decimal does not leave a remainder of zero, but repeats the digit or digits in the quotient, divide to the thousandths' place and round to the nearest hundredth. Such a decimal is called a **repeating decimal**.

$$\frac{7}{9} = 9\overline{)7.000} \quad \frac{0.777 = 0.78}{}$$

$$\begin{array}{r} \underline{63} \\ 70 \\ \underline{63} \\ 70 \\ \underline{63} \\ 7 \end{array}$$

Convert a mixed fraction to an improper fraction before changing it to a decimal.

EXAMPLE

$$1\frac{7}{8} = \frac{15}{8} = 8\overline{)15.000} \quad \frac{1.875 = 1.88}{}$$

$$\begin{array}{r} \underline{8} \\ 70 \\ \underline{64} \\ 60 \\ \underline{56} \\ 40 \\ \underline{40} \end{array}$$

Self-Check

Directions: Convert the following fractions to decimals. Then compare your answers with those in the back of the book.

1. $\frac{5}{6}$ 2. $\frac{11}{12}$ 3. $7\frac{2}{3}$

4. $4\frac{3}{5}$ 5. $\frac{30}{25}$ 6. $9\frac{3}{23}$

Practice

Directions: Convert the following fractions to decimals. Then compare your answers with those in the back of the book.

1. $\frac{3}{8}$ 2. $\frac{9}{10}$ 3. $1\frac{1}{6}$

4. $\frac{8}{15}$ 5. $\frac{5}{12}$ 6. $3\frac{5}{15}$

7. $15\frac{1}{5}$ 8. $\frac{2}{3}$ 9. $2\frac{5}{7}$

Using a Calculator: *Converting Fractions to Decimals*

A calculator converts fractions to decimals very quickly. To convert $\frac{3}{4}$ to a decimal, press **3 ÷ 4 =**. The result is .75. As you can see, the decimal point is already in the right place!

Another nice feature of a calculator is a **repeating decimal**. The calculator does all the work of repeating the numbers. For instance, to convert $\frac{9}{11}$ to a decimal using a calculator with an eight-digit display, the answer will appear as 0.8181818. Rounded to the nearest hundredths' place, your answer will be 0.82.

Lynn Garcia uses a calculator to determine how much fabric she has sold. Convert the amounts to decimals, add them, then round to the nearest hundredth. (Answers are printed below.)

Fabric	Yards	Decimal
Black wool	$15\frac{1}{2}$	_____
Beige/red plaid	$7\frac{1}{8}$	_____
Royal blue silk	$8\frac{1}{3}$	_____
Navy worsted	$10\frac{7}{8}$	_____
Green linen	$6\frac{1}{4}$	_____
Peach cotton	$22\frac{5}{8}$	_____
	Total	

ANSWERS: 15.50, 7.13, 8.33, 10.88, 6.25, 22.63, 70.72

Converting Decimals to Fractions

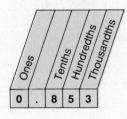

.853
Decimals

Decimal numbers can also be written as fractions. In converting decimals to fractions, you first need to identify the place value of the decimal. In the place value chart, the decimal .8 is in the tenths' place. The decimal is read as eight tenths. To change the decimal to a fraction, place 8 over 10, which is the denominator in the fraction. Here is an example:

$$.8 = \frac{8}{10} \text{ denominator}$$

Note that the number of zeros in the denominator is the same as the number of places to the right of the decimal point. Let's look at another example:

$$.049 = \frac{49}{1000} \text{ (forty-nine thousandths)}$$

After changing a decimal to a fraction, the fraction is reduced to its lowest common denominator, for example:

$$.125 = \frac{125}{1000} = \frac{1}{8}$$

When a mixed decimal (a whole number and a decimal) is converted to a fraction, reduce the decimal only to its lowest common denominator and let the whole number stand. Let's look at two examples:

$$10.25 = 10\frac{25}{100} = 10\frac{1}{4}$$
$$2.5 = 2\frac{5}{10} = 2\frac{1}{2}$$

Self-Check

Directions: Change the mixed decimals and decimals that follow to fractions and reduce to the lowest common denominator. Then compare your answers with those in the back of the book.

1. .05	2. 16.0004	3. 2.69	4. .2875	5. .9
6. 1.065	7. .175	8. .25	9. .73	10. .888

Practice

Directions: Change the decimals to fractions and reduce to lowest terms. Then compare your answers with those in the back of the book.

1. .37	2. 4.512	3. 2.5
4. 11.275	5. 9.654	6. 11.05
7. .311	8. 3.606	9. 24.345

Help with *Aliquot Parts*

Sharon Cartwright works in an auto parts store. She uses aliquot parts to quickly determine prices. An aliquot part is any number that can be divided into another number with no remainder. For example, aliquot parts of $1.00 are $.50 ($\frac{1}{2}$ of $1.00); $.25 ($\frac{1}{4}$ of $1.00); $.20 ($\frac{1}{5}$ of $1.00), and so forth. Multiples of aliquot parts, such as $.75 ($\frac{3}{4}$ of $1.00) can also be treated as aliquot parts. Although any number can be an aliquot part, aliquot parts of 1 are used most since they relate to amounts that can be changed to fractional parts of 100, 100 percent, or $1.00.

Sharon has an invoice for 800 bottles of windshield washer fluid that cost $.375 each. She knows that the aliquot part of $.375 is $\frac{3}{8}$. Mentally she can calculate $\frac{3}{8} \times 800$ and come up with the price of $300.

Learning the aliquot parts of 100 in the following table can help you solve problems quickly and easily.

$25 = \frac{1}{4}$	$20 = \frac{1}{5}$	$12\frac{1}{2} = \frac{1}{8}$	$33\frac{1}{3} = \frac{1}{3}$
$50 = \frac{1}{2}$	$40 = \frac{2}{5}$	$37\frac{1}{2} = \frac{3}{8}$	$66\frac{2}{3} = \frac{2}{3}$
$75 = \frac{3}{4}$	$60 = \frac{3}{5}$	$62\frac{1}{2} = \frac{5}{8}$	$16\frac{2}{3} = \frac{1}{6}$
	$80 = \frac{4}{5}$	$87\frac{1}{2} = \frac{7}{8}$	$83\frac{1}{3} = \frac{5}{6}$

Unit 3 Review

Developing Your Skills

Directions: Follow the instructions for each set of fraction problems. Complete these sets of fractions so that they are equivalent fractions.

1. $\frac{12}{20} = \frac{?}{240}$

2. $\frac{3}{9} = \frac{27}{?}$

3. $\frac{64}{72} = \frac{?}{9}$

4. $\frac{84}{210} = \frac{?}{5}$

5. $\frac{11}{25} = \frac{33}{?}$

6. $\frac{2}{18} = \frac{?}{288}$

Reduce these fractions to their lowest terms. If the fraction cannot be reduced, circle it.

7. $\frac{10}{24}$

8. $\frac{28}{7}$

9. $\frac{49}{84}$

10. $\frac{34}{54}$

11. $\frac{22}{4}$

12. $\frac{13}{26}$

13. $\frac{18}{39}$

14. $\frac{21}{43}$

15. $\frac{52}{8}$

16. $\frac{34}{85}$

Convert the following mixed fractions to improper fractions.

17. $15\frac{9}{34}$

18. $24\frac{7}{27}$

19. $34\frac{5}{15}$

20. $5\frac{3}{5}$

21. $35\frac{9}{84}$

22. $125\frac{3}{10}$

Add the following fractions. Reduce to lowest terms.

23. $\frac{12}{24} + \frac{5}{24}$

24. $\frac{8}{30} + \frac{3}{15}$

25. $\frac{7}{8} + \frac{14}{16}$

26. $4\frac{2}{6} + 5\frac{9}{10}$

27. $5\frac{9}{2} + 10\frac{3}{8}$

28. $3\frac{1}{2} + 4\frac{3}{4} + 6\frac{7}{8}$

Subtract the following fractions. Reduce to lowest terms.

29. $\frac{21}{12} - \frac{7}{6}$

30. $135\frac{16}{48} - 116\frac{6}{16}$

31. $\frac{40}{21} - \frac{60}{42}$

32. $5\frac{3}{6} - \frac{4}{2}$

33. $12\frac{3}{22} - \frac{4}{2}$

34. $72\frac{8}{16} - \frac{9}{4}$

Multiply the following fractions. Reduce to lowest terms.

35. $2\frac{5}{6} \times 6\frac{8}{9}$ 36. $13\frac{6}{12} \times 2\frac{4}{8}$ 37. $\frac{6}{15} \times \frac{9}{30}$

38. $\frac{4}{5} \times \frac{3}{4}$ 39. $\frac{11}{22} \times \frac{1}{2}$ 40. $15 \times 3\frac{5}{9}$

Divide the following fractions. Reduce to lowest terms.

41. $\frac{3}{4} \div \frac{9}{12}$ 42. $\frac{12}{64} \div \frac{3}{16}$ 43. $\frac{4}{5} \div \frac{1}{3}$

44. $\frac{7}{8} \div \frac{5}{7}$ 45. $2\frac{1}{8} \div 5\frac{19}{34}$ 46. $\frac{7}{16} \div \frac{21}{48}$

47. $\frac{3}{10} \div 5\frac{1}{5}$ 48. $3\frac{3}{5} \div 5\frac{1}{4}$ 49. $\frac{19}{23} \div \frac{1}{5}$

Solve the following fraction problems. Reduce to lowest terms.

50. $\frac{5}{8} + \frac{3}{4}$ 51. $4\frac{3}{7} + 6\frac{4}{7}$ 52. $17\frac{11}{28} - 15\frac{2}{7}$

53. $\frac{5}{9} \times \frac{3}{5}$ 54. $\frac{12}{15} \times \frac{3}{4}$ 55. $\frac{1}{16} + \frac{3}{4} + \frac{11}{12}$

56. $11\frac{5}{8} \times 5\frac{1}{3}$ 57. $4\frac{1}{8} - 3\frac{5}{8}$ 58. $3\frac{5}{9} + 5\frac{1}{3} + 4\frac{1}{6}$

59. $6\frac{2}{7} \div 4\frac{4}{14}$ 60. $\frac{7}{54} \times \frac{9}{21}$ 61. $16\frac{5}{12} - 11\frac{2}{3}$

Convert the following fractions to decimals.

62. $\frac{36}{12}$ 63. $\frac{90}{9}$ 64. $\frac{5}{6}$ 65. $\frac{50}{25}$

66. $\frac{19}{29}$ 67. $\frac{1}{7}$ 68. $\frac{80}{6}$ 69. $\frac{120}{350}$

Convert the following decimals to fractions. Reduce to lowest terms.

70. .625 _____

71. .004 _____

72. 2.75 _____

73. .175 _____

74. .85 _____

Working with Percents

PERCENT GROWTH BY YEAR

INTEREST RATES 8.7% 9.67% 10.25%

A jewelry store advertises 20 percent off all merchandise. An automobile dealer talks about the growing percentage of senior citizens driving automobiles. For a percent to have meaning, it must be expressed in relation to a whole. Knowing what percent means and how to figure problems with percents is very useful, especially in the business world.

Understanding Percents

Percent and the percent symbol (%) mean *per hundred*. Therefore, any percent can be converted to a fraction with a denominator of 100. A percent can also be converted to a decimal in the hundredths' place. So 68% can be stated two ways:

1. As a fraction $\frac{68}{100}$

2. As a decimal .68 (68 hundredths)

Remember that in word problems, there may be several ways of saying something. For example, the phrase "sixty-five out of every one hundred students work" can be also stated as "65% of the students work."

Self-Check

Directions: Rewrite the following phrases using percents. Then compare your answers with those in the back of the book.

1. $\frac{1}{100}$ *100%*

2. 12 children out of every 100   *12%*

3. Five hundredths   *5%*

4. 69 per 100 tested   *69%*

5. $26 per $100   *26%*

6. 1 taxpayer out of every 100   *1%*

Practice

Directions: Rewrite the following phrases using percents. Then compare your answers with those in the back of the book.

1. .23   *23%*

2. Forty hundredths   *40%*

3. $\frac{75}{100}$ *75%*

4. 15 athletes out of every 100   *15%*

5. $\frac{100}{100}$ *100%*

6. $\frac{25}{100}$ *25%*

Converting Percents

Of course, as you may have noted in the examples shown earlier, a percent has very little meaning by itself. For instance, an automotive store may offer 10% off. But off of what? The retail price? The manufacturer's suggested price? A special sale price? Before you rush out to buy that item, you may want to know how much you're actually saving. To do this, first you'll have to relate the percent saving to some other number — in this case a price. Then you'll have to convert the percent saving to a dollar amount.

Converting Decimals to Percents

To convert a decimal to a percent, simply move the decimal point two places to the right and add the percent sign (%). So, to convert .2 to a percent, do the following:

Step 1   Move the decimal point two places to the right.   .20

Step 2   Drop the decimal point and add the percent sign.   20%

When converting decimals in the thousandths' place, you follow the same steps. Let's convert .242 to a percent.

Step 1   Move the decimal point two places to the right.   .24,2

Step 2   Add the percent sign.   24.2%

Self-Check

Directions: Convert the following decimals to percents. Then compare your answers to those in the back of the book.

1. .404   40.4%
2. .2   20.%
3. .342   34.2%
4. .06   6%
5. 3.33   333%
6. .09   9%
7. .005   .5%
8. .12   12%
9. 34.332   3433.2%

Converting Percents to Decimals

To convert a percent to a decimal, first drop the percent sign, then divide by 100. A simple way to divide by 100 is to move the decimal point two places to the left. Remember that all whole numbers have a decimal point, even if it is not shown. Convert 80% to a decimal as follows:

Step 1   Drop the percent sign.   80%

Step 2   Move the decimal point over two places to the left.   .80

Self-Check

Directions: Convert the following percents to decimals. Then compare your answers to those in the back of the book.

1. 56%   .56
2. 35%   .35
3. 21%   .21
4. 15.07%   .1507
5. 99.9%   .999
6. 75.2%   .752
7. 33.33%   .3333
8. 50%   .50
9. .09%   .0009

Converting Fractions to Percents

To convert a fraction to a percent, first change the fraction to a decimal. Then convert the decimal to a percent. In this example, $\frac{1}{2}$ is converted to a percent.

Step 1   Change the fraction into a decimal. Divide the numerator by the denominator. $\frac{1}{2} = 2\overline{)1.0}$ .5 = .5

Step 2   Move the decimal point over two places to the right and drop it. Add the percent sign.   .5 = 50%

When converting fractions to percents, it may be necessary to carry the quotient to more than one place. The fraction $\frac{2}{3}$ is converted to a percent as follows:

Step 1	Divide the numerator by the denominator. Carry the quotient to 5 places.	$\begin{array}{r} 0.66666 \\ 3\overline{)2.00000} \end{array}$
Step 2	Move the decimal point two places to the right and add the percent sign.	$.66666 = 66.666\%$
Step 3	Round to the hundredths' place.	$66.666\% = 66.67\%$

Self-Check

Directions: Convert the following fractions to percents. When necessary, carry quotients to 5 places; round percents to two decimal places. Then compare your answers to those in the back of the book.

1. $\frac{3}{5}$

2. $\frac{11}{25}$

3. $\frac{13}{52}$

4. $\frac{17}{34}$

5. $\frac{3}{15}$

6. $\frac{1}{8}$

Converting Percents to Fractions

To convert a percent to a fraction, drop the percent sign and write the percent as a fraction of the number 100. Reduce to lowest terms. Let's convert 25% to a fraction.

Step 1	Drop the percent sign.	25%
Step 2	Convert to a fraction of 100.	$\frac{25}{100}$
Step 3	Reduce to lowest terms.	$\frac{25}{100} = \frac{1}{4}$

Sometimes the number you want to convert may not be expressed as a whole number. For example, suppose you want to convert 38.5% to a fraction. You can still get a whole number in the numerator by (a) multiplying the numerator by 10, 100, 1,000, or whatever factor of 10 is necessary to eliminate the decimal and (b) multiplying the denominator by the same number. To convert 38.5% to a fraction, you do the following:

Step 1	Multiply both the numerator and denominator by 10.	$\frac{38.5}{100} \times \frac{10}{10} = \frac{385}{1,000}$
Step 2	Reduce to lowest terms.	$\frac{385}{1,000} = \frac{77}{200}$

Self-Check

Directions: Convert the following percents to fractions. Then compare your answers with those in the back of the book.

1. 50%

2. 34%

3. 65%

4. 43.5%

5. 10.4%

6. 75%

7. 75.6%

8. 55.5%

9. 11.1%

Practice

Directions: Convert the following problems using the **appropriate steps.** Then compare your answers with those in the back of the book.

Convert to percents.

1. 0.006	2. 0.15	3. 1.755
4. 0.67	5. 3.001	6. 0.7563
7. 6.75	8. 0.0507	9. 3.75

Convert to decimals.

10. 17.5%	11. 115.2%	12. 50%
13. 7.9%	14. 317.65%	15. 0.45%
16. 54%	17. 513%	18. 0.9%

Convert to percents.

19. $\frac{7}{9}$	20. $\frac{3}{7}$	21. $\frac{5}{11}$
22. $\frac{7}{8}$	23. $\frac{4}{8}$	24. $\frac{4}{12}$

Convert to fractions.

25. 20%	26. 50%	27. 10%
28. 75%	29. 60%	30. 33%
31. 48.5%	32. 28.6%	33. 88.2%

Section 3

Using Formulas in Percentage Problems

In mathematics, formulas are often used to help solve problems. A **formula** is simply a form of mathematical shorthand in which symbols—usually letters and operation signs—are used to express relationships. When the number values of several of the letter symbols are known, the unknown values can be determined.

Suppose you wanted to find the area of the top of your desk. The area of any rectangle is determined by multiplying its length by its width. This relationship can be expressed in the formula $A = L \times W$, where A stands for area, L stands for length, and W stands for width. So, if the length (L) of your desktop was 2 feet, and the width (W) was 3 feet, the area (A) would be 6 feet ($A = 2$ ft. $\times$ 3 ft. $= 6$ sq. ft.).

Sometimes the terms *percent* and *percentage* are used as if they mean exactly the same thing. However, **percent**—as explained earlier—means per hundred. It is the rate at which something is changing. **Percentage** means a proportion or part of something in relation to its whole. It refers to the actual amount of the change, which is the product that results when a given number is multiplied by a given percent.

You can find a percentage of a number by using the formula $P = B \times R$. In this formula, P = percentage, B = the base or given number, and R = the rate or given percent. To see how the percentage formula is applied, look at this example.

EXAMPLE

The Peterson Company is giving a cost-of-living pay increase at the rate of 6% to its employees. If Michelle Portier's salary is $15,000, what would be the amount of her increase? To determine Michelle's pay increase, use the percentage formula: $P = B \times R$

Michelle's base salary is $15,000.	$B = \$15,000$
The rate of increase is 6%.	$R = 6\% = .06$
Multiply ($B \times R$) to get the amount of increase (P).	$P = \$15,000 \times .06$ $P = \$900$

Here is another example. What is $16\frac{3}{4}\%$ of 800? Use the formula, $P = B \times R$.

Multiply ($B \times R$) to get the percentage of the whole (P).	$B = 800$ $R = 16\frac{3}{4}\%$ or .1675 $P = 800 \times .1675 = 134$

Self-Check

Directions: Find the percentage in each of the following problems. Then compare your answers to those in the back of the book.

1. 60% of $129
2. 8% of $14.25
3. 14% of 7.50

4. 14% of $2
5. 140% of 610
6. $\frac{1}{2}\%$ of 30

Up to now, you have used the percentage formula only to solve for P. Remember, though, that when any two parts of a formula are known, you can find the unknown part. Therefore, you can easily solve for the remaining part. For example:

$$R = P \div B$$
$$B = P \div R$$

Self-Check

Directions: Use the percentage formula to solve the following problems. Then compare your answers with those in the back of the book.

Find the percentage in each of the following problems.

1. _____% of 100 is 40
2. _____% of 75 is 30

3. _____% of 112 is 35
4. _____% of 80 is 48

Find the base number in each of the following problems.

5. 147 is 35% of _____   6. 63 is 12% of _____

7. $255 is 60% of _____   8. 1,785 is 105% of _____

Practice

Directions: Use the percentage formula to solve the following problems. Then compare your answers with those in the back of the book.

Determine the percentages for the following problems.

1. 25% of $750 is _____   2. 5% of $.77 is _____

3. _____ is 14% of $14.25   4. 35.5% of 2,600 is _____

5. $19\frac{1}{2}$% of 4,000 is _____   6. _____ is 20% of 160

Determine the rates for the following problems.

7. _____% of 300 is 45   8. _____% of $150.00 is $8.25

9. _____% of 800 is 40   10. _____% of 2,000 is 2,000

11. 13 is _____% of 65   12. What percent of 176 is 440?

Determine the base numbers for the following problems.

13. 4% of _____ is 120   14. 60 is 120% of _____

15. 3% of _____ is 15.06   16. 440 is 16% of _____

17. 15% of _____ is 45   18. 128 is 80% of _____

Using a Calculator: *Percents*

If your calculator has a percent key, here's how to use it. Let's say you want to know what 70% of 56 is. Press the following keys: **5 6 × 7 0 %**. The answer is 39.2. You will notice that pressing the percent key accomplishes the same thing as pressing the equals key. To find 35% of 800, press these keys: **8 0 0 × 3 5 %**. The answer is 280.

Jenny McRae finds that using a calculator is faster than looking up the amount of the sales tax on a table and adding it to the cost of an item to determine the price. For instance, the state tax is 5%. She sells a roll of film that costs $3.98. Using her calculator, she can quickly determine the total price in one of two ways. One way is to press **3 . 9 8 + 5 %**. The answer is 4.179. Or she can press **3 . 9 8 × 1 0 5 %**. Multiplying by 105% of the cost is the same as adding the sales tax to the cost. The answer is 4.179. The total cost, rounded to the nearest cent, is $4.18.

Ratios and Rates

Ratios and rates are both comparisons. A *ratio* in math is the comparison of two quantities expressed as the quotient of one divided by the other. A *rate* is the comparison of a quantity measured with respect to another measured quantity. Look at these two types of comparisons.

Ratios

A **ratio** is a comparison of two numbers or quantities expressed by division. For example, a certain factory has 4 out of every 10 products rejected by its quality control department. Although the relationship can be stated as a fraction ($\frac{4}{10}$), it is most frequently stated as follows:

4 to 10 or 4:10

Ratios are written in lowest terms. Reduced to its lowest terms, 4:10 = 2:5.

Let's look at another example of how ratios are used. Each month a sports equipment store sells 250 pairs of white shoes and 75 pairs of red shoes. What is the ratio of white to red shoes?

Step 1 The ratio can be stated either way. $250 \text{ to } 75 = \frac{250}{75}$

Step 2 Reduce to lowest terms. $\frac{250}{75} = \frac{10}{3}$

Self-Check

Directions: Reduce the following ratios to lowest terms. Show your answers the same way the problem is stated. Then compare your answers with those in the back of the book.

1. 5:10
2. 350 to 50
3. 25:75
4. 300 to 30
5. 2,500:50
6. 1,200 to 12

Rates

A **rate** is a ratio that is used to compare quantities of different kinds. Rates are usually written as a ratio of a quantity to 1, called the unit rate. Miles per hour is one frequently used form of unit rate. Cents per pound is another familiar form of unit rate. Look at the following examples.

George and Beth drove 200 miles in 4 hours. How many miles per hour did they drive?

Step 1 Read the problem as a fraction. $\frac{200}{4}$

Step 2 Divide 200 miles by 4 hours. $4\overline{)200}$

Step 3 Miles per hour equals 50. $4\overline{)200}^{\,50}$

George and Beth drove at a rate of 50 miles per hour.

Here is another example.

In a supermarket, a unit price is a frequently used form of unit rate. If a unit equals one pound, then find the unit price of a 2-lb. package of dried peas that costs $.68.

Step 1   A 2-pound package is $.68. Divide $.68 by 2. $2\overline{)\$.68}$

Step 2   One pound costs $.34. The unit price is $.34 per pound. $2\overline{)\$.68}^{\$.34}$

Self-Check

Directions: Find the unit rates. Then compare your answers with those in the back of the book.

1. Walking 5 miles in 2 hours _____

2. Traveling 600 miles in 4 hours _____

3. Processing 24 photos for $8.40 _____

4. Skiing 20.1 miles in 3 hours _____

5. Receiving $15.40 for 4 hours of work _____

Practice

Directions: Follow the instructions for each part of this exercise. Then compare your answers with those in the back of the book.

Jane is planning a business luncheon. Use the information from the responses to write the ratio for each problem. Reduce ratios to their lowest terms.

Food Choices	Number of Responses
Beef	240
Chicken	300
Pork	120
Seafood	220
Vegetarian	70

1. Ratio of vegetarian to pork _____

2. Ratio of chicken to seafood _____

3. Ratio of chicken to beef _____

4. Ratio of pork to beef _____

5. Ratio of vegetarian to beef _____

Find the rates for the following problems.

6. Driving 200 miles in 4 hours _____

7. Paying $4.05 for $7\frac{1}{2}$ pounds of tomatoes _____

8. Buying 10 ounces of soda for $.75 _____

9. Buying 3 pounds of grapes for $1.29 _____

10. Buying 5 pounds of bananas for $1.09 _____

Use the percentage formula to identify the rate (*R*), base (*B*), or percentage (*P*).

11. 6% of 33 is _____

12. _____ is 25% of 40

13. 20% of _____ is 23

14. $500 is 50% of _____

15. _____% of $3,000 is $375

Help with *Unit Prices*

Unit prices may seem like a mystery to you. However, a wise shopper uses unit prices to help get the best buy. In the supermarket, you will find that cereals come in all sizes: family, economy, king, and giant-size. Boxes of cereal also come in varying sizes by weight: 10 oz., 18 oz., 24 oz., and so forth. How can you tell which is the best buy? Look at the unit prices. For example:

Cereal	Size	Price	Unit Price
Vigorite	24 oz.	$2.63	$1.75 per lb.
Toasties	1 lb. 6 oz.	$2.49	$1.81 per lb.
Oats Plus	2 lb. 2 oz.	$3.36	$1.58 per lb.

What's the best buy? Oats Plus is the best buy at $1.58 per pound. Usually, larger quantities cost less. That is not always the case, though, so check the unit prices to be sure you are getting the best buy.

Using unit prices to help you shop can result in real savings. This is particularly true with products where the quality is much the same. When you see shelves full of similar products, you might let the unit prices help you decide what to buy.

Unit 4 Review

Developing Your Skills

Directions: Follow the instructions for each set of problems.
Convert the following fractions to percents.

1. $\frac{72}{150}$ 2. $\frac{65}{250}$ 3. $\frac{13}{16}$

4. $\frac{16}{100}$ 5. $\frac{46}{50}$ 6. $\frac{28}{400}$

Convert the following decimals to percents.

7. .68   8. 52.66   9. 8.724

10. .74   11. .0825   12. .189

Use the percentage formula to solve the following:

13. 12% of 40 is _____   14. 125% of 270 is _____

15. 3% of 78 is _____   16. .6% of 90 is _____

17. 27% of 60 is _____   18. .68% of 69 is _____

Solve the following:

19.   189 is 9% of what number? _____

20.   499 is 12% of what number? _____

21.   15 is .5% of what number? _____

22.   24 is .5% of what number? _____

23. $3,000.00 is 275% of what number? _____

24. What is 4% of 9? _____   25. What is 75% of 325? _____

26. What is 13% of 65? _____   27. What is 36.45% of 174? _____

28. What is 50% of 350? _____   29. What is 9% of 5? _____

30. 600 is 60% of _____   31. 13 is _____% of 65

32. 125% of 270 is _____   33. 18 is _____% of 30

34. 24 is _____% of 25   35. 3% of 78 is _____

36. 54 is _____% of 675   37. 42.5 is _____% of 85

Reduce the following ratios to lowest terms.

38. 10 to 50

39. 100 to 300

40. 5 to 1,000

41. 400 to 40,000

42. 1 to 22

43. 500 to 50

44. 6,000 to 8

45. 20 to 1

46. 3,000 to 30

Solving Word Problems

Directions: In solving a mathematics word problem, ask: (1) What information is being asked for? (2) What information is given in the problem? (3) What information is needed to solve the problem? Use the formulas in this unit to solve these problems.

47. Lucia Gomez took a civil service exam for a government job. On one section of the exam she answered 60 of 75 questions correctly. What percent of the questions did she get right? _____

48. Tony's Record Store had total sales on a Saturday of $1,655. The top salesperson that day ended with sales of $331. What percent of the total sales did that salesperson have? _____

49. The owners of the stadium knew that approximately $12\frac{1}{2}\%$ of the people who went to a game would buy programs. Last Saturday, attendance at the game was 57,128. Approximately how many programs were sold? _____

50. The owners of the stadium know that gross profit from a rock concert will be 50% of ticket sales of $75,000. What will be the gross profit? _____

Income

Income is money that a person receives from a variety of sources, such as a job, savings interest, or investments. If you work for a large company, your supervisor and the personnel department usually determine your pay, based on past information that you supply. The payroll department calculates the exact amount of your paycheck. It is a good idea, however, to know exactly how this amount is determined. If a mistake is made, you'll know it!

Determining Gross Pay

The total amount of money that a person earns from a job is called **gross pay**. Gross pay can be computed in many ways. Common methods of figuring gross pay are straight-time pay, overtime pay, piecework pay, salary, and commission pay.

Straight-Time Pay

Many people earn an hourly rate of pay. The **hourly rate** is a set amount of money paid for each hour worked. For example, a person may have an hourly rate of $5.00 per hour or $7.25 per hour. Gross pay that is based on an hourly rate of pay and the number of hours worked is called **straight-time pay**. Here is the formula for straight-time pay.

Hourly Rate × Hours Worked = Straight-Time Pay

EXAMPLE

Mike Black is paid $10.25 per hour. What would Mike's straight-time pay be if he worked $30\frac{1}{2}$ hours last week?

Hourly Rate × Hours Worked = Straight-Time Pay

$10.25 × 30.5   = $312.625 or $312.63

Self-Check

Directions: Determine the straight-time pay in the following situations. Round answers to the nearest cent. Then compare your answers with those in the back of the book.

1. Pedro Jones worked 24 hours last week for a nursery. He is paid $4.85 per hour. What was his straight-time pay? _____

2. Jere Washington, a cashier, earns $4.40 per hour. Last week she worked 18 hours. What was her straight-time pay? _____

3. John Laskowitz is a mechanic at a bicycle shop. He earns $6.15 per hour. Last week he worked $37\frac{1}{2}$ hours. What was his straight-time pay? _____

4. Larry Crisp has a part-time job at a supermarket. He earns $4.55 per hour. Last week he worked $16\frac{1}{4}$ hours. What was his straight-time pay? _____

5. At the paper mill where she works, Evelyn Amori has a $7\frac{1}{2}$-hour shift each day. She works 5 days a week and earns $6.20 an hour. What is her straight-time pay for the week? _____

Overtime Pay

Work beyond regular hours, called **overtime**, is paid for at a different rate. Often it is paid at a rate of **time-and-a-half**, or $1\frac{1}{2}$ times the regular hourly rate. Sometimes work on holidays or Sundays is paid for at a rate of **double time**, or two times the regular hourly rate. The rates of pay, time-and-a-half and double time, are overtime factors. Here are the formulas for calculating gross pay with overtime.

Hourly Rate × Overtime Factor	= Overtime Hourly Rate
Overtime Hourly Rate × Overtime Hours Worked	= Overtime Pay
Straight-Time Pay + Overtime Pay	= Gross Pay

EXAMPLE

Katie Lewandowski has a job as a data-entry operator. She earns $7.15 per hour for a regular 40-hour week. She is paid time-and-a-half for working on Saturdays. She is paid double time for working on Sundays. Last week she worked her regular 40-hour week plus 8 hours on Saturday and $3\frac{1}{2}$ hours on Sunday. What was her gross pay?

Step 1   Calculate straight-time pay.

Hourly Rate × Hours Worked = Straight-Time Pay
$7.15 × 40 = $286.00

Step 2   Calculate time-and-a-half pay.

Hourly Rate × Overtime Factor = Overtime Hourly Rate
$7.15 × 1.5 = $10.725 = $10.73
Overtime Hourly Rate × Overtime Hours Worked = Time-and-a-Half Pay
10.73 × 8 = $85.84

Step 3   Calculate double-time pay.

Hourly Rate × Overtime Factor = Overtime Hourly Rate
$7.15 × 2 = $14.30
Overtime Hourly Rate × Overtime Hours Worked = Double-Time Pay
$14.30 × 3.5 = $50.05

Step 4   Calculate Gross Pay.

Straight-Time Pay + Overtime Pay = Gross Pay
$286.00 + $85.84 + $50.05 = $421.89

Self-Check

Directions: Determine the gross pay in the following situations involving straight-time pay plus overtime. Then compare your answers with those in the back of the book.

1. Jerry Lawrence, a librarian, is paid $7.00 an hour for a 35-hour week. If he works more than 35 hours, he is paid time-and-a-half for overtime. Last week he worked 43 hours. What was his total gross pay? _____

2. Jenny Orlando, a dance instructor, worked 45 hours last week. She is paid $12.00 per hour for a 40-hour week. She is paid time-and-a-half for hours over 40. What was her total gross pay?

_____

3. Rita Perez, a salesclerk, is paid at an hourly rate of $4.50. She is paid time-and-a-half for working on Saturdays. Last week she worked 36 hours during the week and $4\frac{1}{2}$ hours on Saturday. What was her total gross pay? _____

4. Linda Collier, a machinist, works a 40-hour week at a rate of $8.70 per hour. Last week she worked a total of 45 hours, plus 3 hours on Sunday. She is paid time-and-a-half for hours over 40 and double time for working on Sunday. What was her total gross pay for the week? _____

5. Ed Mulaney, a nurse's aide, works a 35-hour week at a rate of $5.45 per hour. He is paid time-and-a-half for overtime. He is paid double time for Sundays. Last week he worked 6 hours overtime and 5 hours on Sunday. What was his total gross pay for the week?

_____

Piecework Pay

Working on a **piecework** basis means you are paid a set amount for each piece of work you complete. Here is the formula for determining gross pay on a piecework basis.

Rate per Piece × Number Produced = Gross Pay

EXAMPLE   Ernie Gotschalk is an assembler. He is paid $1.10 for each part he assembles. Last week he assembled 463 parts. What was his gross pay?

Rate per Piece × Number Produced = Gross Pay
$1.10 × 463   = $509.30

Self-Check

Directions: Determine the gross pay for work paid on a piecework basis. Then compare your answers with those in the back of the book.

Rate per Piece	Number Produced	Gross Pay
1. $0.10	5,500	_____
2. $0.25	1,120	_____
3. $0.35	900	_____
4. $1.10	50	_____
5. $0.50	1,400	_____
6. $0.75	1,200	_____

Salary

A **salary** is a fixed amount of money that is paid on a regular basis to an employee. A salary does not depend on the number of hours worked or the amount of work produced. A salary can be paid on a weekly basis, with 52 pay periods a year; on a biweekly basis, with 26 pay periods a year; on a semimonthly basis, with 24 pay periods a year; or on a monthly basis, with 12 pay periods a year. Gross pay per pay period for salaried employees can be calculated by using the following formula:

Annual Salary ÷ Number of Pay Periods = Gross Pay per Pay Period

EXAMPLE

Jeff Snowberger earns a salary of $24,000 annually. He is paid semimonthly. What is his semimonthly gross pay?

Annual Salary ÷ Number of Pay Periods = Gross Pay per Pay Period
$24,000 ÷ 24　　　　　　　　　　　= $1,000

Here is another formula to calculate gross pay for salaried employees.

Gross Pay per Period × Number of Pay Periods = Annual Salary

EXAMPLE

Patti Watkins has an annual salary. Her biweekly gross pay is $725. What is her annual salary?

Gross Pay per Pay Period × Number of Pay Periods = Annual Salary
$725 × 26　　　　　　　　　　　　　　= $18,850

Self-Check

Directions: Determine the annual salary in the following problems. Then compare your answers with those in the back of the book.

1. Victor Lange is paid weekly. Each week his gross pay is $600. What is his annual salary? _____

2. Paul Criscione makes $15,000 per year. He is paid on a biweekly basis. What is his gross pay per pay period? _____

3. Veronica Lewis has an annual salary of $125,000. She is paid monthly. What is her gross pay per pay period? _____

4. Maria Lopes is paid semimonthly. Her semimonthly gross pay is $1,500. What is her annual income? _____

5. Tiffany Jackson works part-time and makes $100 every week. What is her annual salary? _____

6. Mark Gomez is an insurance adjuster. His annual salary is $18,620. What is his semimonthly salary? _____

Commission Pay

Some workers, particularly salespeople, are paid on a commission basis. A **commission** is pay that is calculated on a percentage of each sale made by the salesperson. The percentage is called a commission rate. If the commission is a salesperson's only pay, it is called **straight commission**. Here is the formula to calculate a commission that is based on a percentage of each sale made.

Commission Rate × Sale Amount = Commission

EXAMPLE   Julia Cadwallader sells pharmaceutical products to hospitals. Her commission rate is 14% of each sale that she makes. Using the above formula, what is her commission on a sale of products totaling $8,560?

Commission Rate × Sale Amount = Commission
14% × $8,560.00   = $1,198.40

Self-Check

Directions: Determine the gross pay for the following problems. Then compare your answers with those in the back of the book.

1. Jim Kinsey is a salesperson for Proclein Products. He is paid 18.5% commission on his gross sales. Last week his sales totaled $3,680. What was his gross pay for that week? _____

2. Rosa Delaney is a sales representative for Merrill Bache. She is paid 15% of the total income she generates for Merrill Bache. Last week her performance generated $8,000. What was her gross pay for that week? _____

3. Phil Lomes is an insurance agent who is paid 12% of the total premiums he writes. Last month Phil's total premiums were $12,900. What was his gross pay for the month? _____

4. Ann Quantro receives a commission of 13.5% on all sales she makes. This week her sales totaled $7,300. What is her gross pay for the week? _____

Commission and Salary

Some people's earnings are determined through a combination of salary and commission. For example, a salesperson may be paid a monthly salary plus a commission on every sale that is made. Others may be paid either a salary or a commission, whichever is greater.

EXAMPLE   Janet Davis sells used cars. She earns a salary of $900 per month or an 8% commission on the cars she sells, whichever is greater. Her sales this month total $12,000. What is her gross pay?

Commission Rate × Sales Amount = Gross Pay on Commission
8% × $12,000   = $960

The commission of $960 is greater than the $900 salary. She is paid the commission of $960.

Self-Check

Directions: Determine the gross pay for the following problems. Then compare your answers with those in the back of the book.

1. May Kim is paid a monthly salary of $600 plus a 12% commission on her monthly sales. Her sales for the month total $17,540. What is her gross pay for the month? _____

2. Carlson Whittier is paid a guaranteed salary of $1,000 per month or a 5% commission on sales, whichever is greater. His sales this month total $38,560. What is his total gross pay for the month?

   _____

3. Taro Troy sells office supplies. Taro receives a monthly salary of $2,000 plus 6% commission on any sales over $50,000 in one month. In June, Taro sold $55,000 worth of office supplies. What was Taro's gross pay for the month of June? _____

4. Carlos Tucker sells auto parts. He is guaranteed a salary of $860 a month or $7\frac{1}{4}$% of his total sales, whichever is greater. October sales totaled $10,987 and November sales totaled $9,825. What was his gross pay for October? _____
   What was his gross pay for November? _____

Graduated Commission

Some commission rates increase as a worker's performance increases. **Graduated commission** pays a different rate of commission for each of several levels of performance. A graduated commission could increase the salary of workers in sales or piecework. For instance, someone paid on a piecework basis may receive $1.00 for each of the first 50 pieces produced, and $1.50 for each piece over 50 produced. Here is the formula to calculate gross pay based on a graduated commission.

Sum of Commission for All Levels of Sales = Graduated Commission

EXAMPLE

Jessica Whitney sells art prints. She gets a 10% commission on the first $2,000 of sales, a 15% commission on the next $2,000, and a 20% commission on all sales over $4,000. If her sales for the month total $5,500, what is her commission for the month?

Step 1 The commission for $2,000 at 10%. 10% × $2,000 = $200

Step 2 The commission for $2,000 at 15%. 15% × $2,000 = $300

Step 3 The commission for $1,500 at 20%. 20% × $1,500 = +$300

                                   Total Graduated Commission = $800

Self-Check

Directions: Determine the gross pay in the following problems. Then compare your answers with those in the back of the book.

1. LaBelle Carter sells textbooks. She is paid a graduated commission. She earns a 10% commission on the first $5,000 of sales, 12% commission on the next $5,000 of sales, and 15% on all sales over $10,000. If her sales for the month total $14,000, what is her total commission? _____

2. Tom Morrison works in a curtain rod factory and gets paid a graduated commission on a piecework basis. Tom is paid $.50 for each of the first 200 pieces he produces during an eight-hour shift, and $1.00 each for any pieces over that. Last week Tom worked eight hours each day and produced the following: Monday, 250 units; Tuesday, 190 units; Wednesday, 340 units; Thursday, 400 units; and Friday, 290 units. What was his gross pay last week?

_____

Practice

Directions: Determine the annual salary or gross pay per period in the following problems. Then compare your answers with those in the back of the book.

1. Tom Brennan worked 24 hours last week, and he is paid $4.00 an hour. What was his gross pay for that period?

_____

2. Mary Kisbeth worked 18 hours last week, and she is paid $6.25 per hour. What was her gross pay for that period?

_____

3. Jon Launder worked 37.5 hours last week, and he is paid $5.50 per hour. What was his gross pay for the week?

_____

4. Larry Dobson has an annual salary of $19,000. He is paid monthly. What is his monthly gross pay?

_____

5. Thomas Jorgenson's commission rate is 12.5%. What is his commission on a sale of $326?

_____

6. Marylyn Lee has an annual salary. Her biweekly gross pay is $682.50. What is her annual salary?

_____

7. Jan Kay is paid a monthly salary of $750 plus a 14% commission on her monthly sales. Her sales for the month total $16,250. What is her gross pay for the month?

_____

8. Lisa Smith is a professional hairdresser. Lisa receives $200 per week salary and 15% per $20 haircut that she performs. Last week she completed 50 haircuts. What was her gross pay for the week?

_____

Using a Calculator: *The Equals (=) Key*

On some calculators, the equals (=) key is also called the **constant** or **repeat** key. You can use the equals key to add the same number over and over. For example, if you press **2 + = = = =**, the result is 10. If you press any number and **+** then **=**, the calculator will add the same number to the total as many times as you press the equals key.

You can also use the equals key to multiply by a constant number. Press **2 × = = = =**. Did you get 32? (2 × 2 = 4 × 2 = 8 × 2 = 16 × 2 = 32)

This procedure may not work on all calculators. Try it with your calculator. If you need to add the same number to a total over and over again or if you need to multiply by a constant number, the equals key can save you a lot of time.

Section 2

Determining Net Pay

If you have ever received a paycheck, you've probably noticed that certain amounts of money were taken out, or deducted, from your gross pay. Some of these deductions are required. For instance, employers are required to make deductions for federal income tax and social security. Some states and cities have income taxes, too. In addition, voluntary deductions may be taken for such things as health and life insurance premiums. Some people even have credit union loans and savings deposits deducted from their gross pay. What remains after all deductions is **net pay**, also called **take-home pay**. Net pay is the amount of money that you actually get and can spend.

Gross Pay − Deductions = Net Pay

Federal Income Tax Federal income tax (FIT), often called **withholding tax**, is usually the largest single deduction. Employers have tax tables that show how much money to withhold from your pay. The amount of your withholding tax depends on your gross income, marital status, and withholding allowances. You can claim one withholding allowance for yourself, and another for your spouse if you are married. You can also claim one allowance for each dependent. A **dependent** is a person who relies on you for financial support.

Look at the federal income tax tables on the following page. There is one table for single persons and one table for married persons. On each table, the range of income is on the left side of the table. The number of withholding allowances is across the top of the table.

EXAMPLE

Sarah Goldblum is a medical records supervisor. Last week she worked 40 hours and was paid at a rate of $8.30 per hour. She is single and claims one withholding allowance. How much federal income tax will be deducted from her gross pay for the week?

Step 1   Compute weekly gross pay.

Hourly Rate × Hours Worked = Weekly Gross Pay

$8.30 × 40   = $332.00

Step 2   Look at the tax table for single persons. Find the range of wages that applies to weekly gross pay. The range is $330 to $340.

Step 3   Find the amount of weekly income tax. Move across the table to the number of withholding allowances claimed. One allowance is claimed. The tax is $48.

EXAMPLE

If, on the other hand, Sarah were married, she would use the table for married persons. If she claimed one withholding allowance for herself and one withholding allowance for her husband, her federal income tax would be calculated as follows:

Step 1   Compute weekly gross pay.

Hourly Rate × Hours Worked = Weekly Gross Pay

$8.30 × 40   = $332.00

Step 2   Look at the tax table for married persons. Find the range of wages that applies to weekly gross pay. Again, the range is $330 to $340.

Step 3   Find the amount of weekly income tax. Move across the table to the number of withholding allowances claimed. Two allowances are claimed. The tax is $34.

SINGLE Persons—WEEKLY Payroll Period

| And the wages are— | | And the number of withholding allowances claimed is— | | | | | | | |
| At least | But less than | 0 | 1 | 2 | 3 | 4 | 5 | 6 | 7 |
		The amount of income tax to be withheld shall be—							
145	150	16	13	10	8	5	2	0	0
150	160	18	14	11	9	6	3	1	0
160	170	19	16	13	10	7	4	2	0
170	180	21	18	14	11	9	6	3	1
180	190	22	19	16	13	10	7	4	2
190	200	24	21	18	14	11	9	6	3
200	210	26	22	19	16	13	10	7	4
210	220	27	24	21	18	14	11	9	6
220	230	29	26	22	19	16	13	10	7
230	240	31	27	24	21	18	14	11	9
240	250	33	29	26	22	19	16	13	10
250	260	35	31	27	24	21	18	14	11
260	270	37	33	29	26	22	19	16	13
270	280	39	35	31	27	24	21	18	14
280	290	41	37	33	29	26	22	19	16
290	300	43	39	35	31	27	24	21	18
300	310	46	41	37	33	29	26	22	19
310	320	48	43	39	35	31	27	24	21
320	330	50	46	41	37	33	29	26	22
330	340	53	48	43	39	35	31	27	24

MARRIED Persons—WEEKLY Payroll Period

| And the wages are— | | And the number of withholding allowances claimed is— | | | | | | | |
| At least | But less than | 0 | 1 | 2 | 3 | 4 | 5 | 6 | 7 |
		The amount of income tax to be withheld shall be—							
200	210	20	17	14	11	9	6	4	2
210	220	21	18	16	13	10	8	5	3
220	230	23	20	17	14	11	9	6	4
230	240	24	21	18	16	13	10	8	5
240	250	26	23	20	17	14	11	9	6
250	260	28	24	21	18	16	13	10	8
260	270	29	26	23	20	17	14	11	9
270	280	31	28	24	21	18	16	13	10
280	290	32	29	26	23	20	17	14	11
290	300	34	31	28	24	21	18	16	13
300	310	36	32	29	26	23	20	17	14
310	320	38	34	31	28	24	21	18	16
320	330	39	36	32	29	26	23	20	17
330	340	41	38	34	31	28	24	21	18
340	350	43	39	36	32	29	26	23	20
350	360	45	41	38	34	31	28	24	21
360	370	47	43	39	36	32	29	26	23
370	380	48	45	41	38	34	31	28	24
380	390	50	47	43	39	36	32	29	26
390	400	52	48	45	41	38	34	31	28
400	410	55	50	47	43	39	36	32	29
410	420	57	52	48	45	41	38	34	31
420	430	59	55	50	47	43	39	36	32
430	440	61	57	52	48	45	41	38	34
440	450	63	59	55	50	47	43	39	36

Self-Check

Directions: Compute each person's gross weekly pay. Use the federal withholding tax tables to determine the amount of federal income tax deducted from each employee's weekly gross pay. Then compare your answers with those in the back of the book.

Name	Hours Worked	Hourly Rate	Weekly Gross Pay	Marital Status	Withholding Allowances Claimed	Income Tax Deducted
1. L. Preston	40	$9.50	_____	Married	3	_____
2. G. Lucci	40	$6.00	_____	Single	1	_____
3. K. Lee	38	$7.35	_____	Single	3	_____
4. M. Kim	37	$8.85	_____	Married	2	_____
5. D. Finn	26	$14.80	_____	Married	5	_____

Social Security Tax

Social security tax, often called **FICA**, is another deduction from your gross pay. FICA is used by the federal government to provide retirement income and hospitalization insurance for people over 65 years of age. FICA also provides benefits to the disabled and benefits to help the dependents of workers who have died.

Unlike federal or state income taxes, FICA taxes are not necessarily applied to all of one's income. Instead, they are deducted from an employee's earnings each pay period until a certain maximum for each year is reached. This maximum may change from year to year. Similarly, the FICA tax rates may change. For our purposes, assume that currently the FICA tax rate is 7.51% on the first $45,000 of income. To determine FICA deductions, use this formula:

FICA Tax Rate × Gross Pay = FICA Deduction

EXAMPLE John Nye has an annual salary of $13,780. He is paid on a biweekly basis. Based on a FICA tax rate of 7.51%, what is his FICA tax per pay period?

Step 1 Find the gross income per pay period.

Annual Pay ÷ Number of Pay Periods = Gross Pay per Period

$13,780.00 ÷ 26   = $530.00

Step 2 Find the deduction per pay period.

FICA Rate × Gross Pay per Period = Deduction per Pay Period

7.51% × $530.00   = $39.80

Self-Check

Directions: Compute the gross pay per period and the amount of FICA tax deducted for each employee. Use the FICA rate of 7.51% for income up to $45,000. Then compare your answers with those in the back of the book.

1. Kai Wang worked 40 hours at $7.85 per hour. How much FICA tax will be deducted from her gross pay?

   _____

2. Jim Goya is paid an annual salary of $30,000. He is paid monthly. How much FICA tax will be deducted from his gross pay each month? _____

3. Terry Esparza is paid 12% straight commission. Last week his sales totaled $20,000. How much FICA tax will be deducted from his gross pay? _____

4. Todd Garcia is paid an annual salary of $54,000. How much FICA tax is deducted from his gross pay each year?

   _____

5. Last week Tanya Pratt worked 40 hours at $8.00 per hour, 5 hours at time-and-a-half, and 6 hours at double time. How much was the FICA deduction from her gross pay?

   _____

Voluntary Deductions

Voluntary deductions are made after federal, state, and other required deductions are taken. **Voluntary deductions** may include savings plans and health, dental, and life insurance. To compute net pay, subtract all required deductions from gross pay, then subtract voluntary deductions.

EXAMPLE

Tom Blake's gross pay last year was $20,904. He is married and claims three withholding allowances for federal income tax. He also has a state income tax that is computed at 2% of gross pay. FICA is 7.51% of gross income. He has a voluntary deduction of $100 per month for his children's college savings. He has an additional monthly deduction of $65 taken for health insurance. What was Tom's net pay last year?

Step 1 Calculate total federal income tax deducted from gross pay.

Gross Annual Pay ÷ Weeks in Year = Gross Weekly Pay

$20,904.00 ÷ 52   = $402.00

Income Tax per Week (from table)   = $43.00

Weeks in Year × Weekly Income Tax = Annual Income Tax

52 × $43.00   = $2,236.00

Step 2 Calculate state income tax.

Gross Income × State Income Tax Rate = State Income Tax

$20,904.00 × 2%   = $418.08

Step 3 Calculate FICA tax withheld.

FICA Tax Rate × Gross Annual Income = Annual FICA Tax

7.51% × $20,904.00   = $1,569.89

Step 4 Compute voluntary deductions.

Monthly Deduction × 12 Months   = Yearly Deduction

$100.00 × 12   = $1,200.00

Monthly Deduction × 12 Months   = Yearly Deduction

$65.00 × 12   = $780.00

Step 5 Add all deductions.

Annual Income Tax	$2,236.00
State Income Tax	418.08
FICA Tax	1,569.89
Savings	1,200.00
Health Insurance	780.00
Total Deductions	$6,203.97

Step 6 Find net pay.

Gross Income − All Deductions   = Net Pay

$20,904.00 − $6,203.97

= $14,700.03

Self-Check

Directions: In each of the following problems, calculate the net pay from the information given. Use the federal income tax tables. Compute FICA at 7.51% for gross income up to $45,000. Then compare your answers with those in the back of the book.

1. Allen Barchus had gross wages of $230.50 last week. He is single and claims one withholding allowance. What was his net pay last week after federal income tax and FICA deductions?

2. Ross Underwood's gross pay for one year was $14,040. He is single and claims one withholding allowance. What was his net pay for the year after federal income tax and FICA deductions?

_____

3. Jackie Lambert's gross pay last week was $345. She is married and claims two withholding allowances. She has a state income tax rate of 1.5% on gross pay. What was her net pay for the week after federal, state, and FICA taxes?

_____

4. Clare Kinsinger's gross pay for one year was $14,357.80. She is single and claims one withholding allowance. She has voluntary deductions of $43.20 per month for health insurance and $15.00 per month for union dues. What was her net pay for the year after federal income tax, FICA, and voluntary deductions?

_____

5. Ellen Kennedy's gross pay is $13,299 per year. She is married and claims 3 withholding allowances. The state income tax rate is 2% of gross pay. Every week she has voluntary deductions of $6.10 for health care, $5.10 for life insurance, and $5.00 for the Christmas savings club. What is her net pay for the year after federal and state income taxes, FICA, and voluntary deductions?

_____

Practice

Directions: Solve the following word problems. Then compare your answers with those in the back of the book.

1. Evelyn Stoneham makes $6.30 an hour as a short-order cook. She is paid time-and-a-half for overtime after 40 hours and double time for Sunday. Her weekly timecard showed these hours:

M	T	W	T	F	S	S
7	8	9	8	8	7	5

What was her gross pay for that week? _____

2. Lisa Johnson has been offered two jobs. Alpha Company offered her a job that pays a weekly salary of $346.16. Beta Industries offered her a job that pays $8.41 per hour for a 40-hour week. How much annual gross pay will she earn at Alpha Company?

_____

How much annual gross pay will she earn at Beta Industries?

_____

3. Using the tax table and a FICA rate of 7.51%, calculate the weekly net pay for Bernie Manns based on the following information. He is paid a weekly salary of $225.00, plus a commission of 5% on all sales for the week. Last week his sales were $3,362.63. He is married and claims two withholding allowances for federal income tax. He has a voluntary deduction of $2.23 a week for health insurance. What is his weekly net pay? _____

4. Julie DeFalco earns $7.50 per hour. She is married and claims two withholding allowances. Last week Julie worked 40 hours at the regular rate and 4 hours at time-and-a-half. Julie has a weekly voluntary deduction of $38.50 for health insurance. Using the tax table and FICA rate of 7.51%, figure Julie's net pay.

Help with *Choosing a Job*

When deciding whether to accept a job, you might want to consider the fringe benefits as well as the salary being offered. **Fringe benefits** are provided by the employer at little or no cost to an employee. These benefits may include such items as health insurance, dental insurance, disability leave, life insurance, vacation time, tuition reimbursement, and retirement plans. Remember, your total compensation includes not only your gross pay, but also the cost of any fringe benefits.

What would you do in the following situation? Company A pays $23,000 a year and offers the following benefits: full payment of health insurance; 80 percent payment of dental insurance; 3 weeks of vacation; short-term disability leave; tuition reimbursement. Company B pays $24,500 a year and offers the following benefits: full payment of health insurance; full payment of dental insurance; 2 weeks of vacation; short- and long-term disability leave.

Young people might consider Company A more favorably because of the vacation time and the opportunity to get an advanced degree at no cost through the tuition reimbursement plan. People with children might consider Company B more favorably because of the dental insurance. Older workers might also look at Company B more favorably because of the long-term disability leave.

As you can see, fringe benefits are just as important as salary in deciding whether to accept a job.

Unit 5 Review

Developing Your Skills

Directions: Use the information given to find gross pay. Round all figures to the nearest cent.

Employee	Hourly Rate	Regular Hours Worked	Overtime Hours Worked	Overtime Pay Rate	Gross Pay
1. La Pierre, E.	$4.25	37	0	$1\frac{1}{2}$	_____
2. Kawano, S.	$6.20	40	0	$1\frac{1}{2}$	_____
3. Majias, P.	$5.85	40	3	2	_____
4. Swenson, M.	$4.50	40	5	$1\frac{1}{2}$	_____

Employee	Piecework Rate	Number Produced	Gross Pay
5. Amber, J.	$.135	430	_____
6. Chan, K.	$.75	190	_____
7. Luis, P.	$.24	386	_____
8. Mouis, M.	$.19	267	_____

Employee	Sales	Commission Rate	Gross Pay
9. Barsack, D.	$5,800.00	$6\frac{3}{4}\%$	_____
10. Gleason, N.	$965.00	9%	_____
11. Palance, J.	$11,675.00	$1\frac{1}{2}\%$	_____
12. Tieso, K.	$1,011.00	$7\frac{1}{4}\%$	_____

The following employees receive either a monthly salary or a commission on their sales, whichever is greater. Find the gross pay for each.

Employee	Salary	Sales	Commission Rate	Gross Pay
13. Albe, C.	$800.00	$9,000.00	10%	_____
14. Flood, J.	$775.00	$5,426.00	17%	_____
15. Maja, J.	$900.00	$4,621.55	21.5%	_____
16. Pons, L.	$700.00	$8,542.21	6.55%	_____

Solving Word Problems

Directions: Use the income tax tables on pages 77 and 79 of this unit and a FICA rate of 7.51% to figure the problems.

17. Lynn Gladowitz is a beautician. She earns an annual salary of $16,470. She is paid on a biweekly basis. What is her gross biweekly pay?

   _____

18. Tyrone Washington is paid a 12% commission on every tuxedo he sells. If Tyrone sells 2 tuxedos for $350.98 each, what is his commission?

   _____

19. Miriam Davis works on an assembly line. She is paid $.11 for every part that she assembles. Last week she assembled 2,880 parts. What was her gross pay for the week?

   _____

20. Lance Von Richter is a travel agent. He is paid either a guaranteed monthly salary of $800 or a 5% commission on the trip packages that he sells, whichever is greater. If Lance sold $45,000 in trip packages last month, what was his gross pay?

   _____

21. Michelle LeRoy has an annual salary of $48,000. Based on a FICA tax rate of 7.51%, how much is deducted every year for FICA taxes?

   _____

22. Luke Deering earns a graduated commission. He earns a 5% commission on the first $3,000 of sales, a 10% commission on the second $3,000 of sales, and a 20% commission on all sales over $6,000. If his sales for the month were $13,500, what was his total commission?

   _____

23. Calculate the weekly net pay for James Brooks based on the following information. James works a 40-hour week and is paid $11.20 an hour. He is married and claims four withholding allowances. The state income tax rate is 1.5% of gross income. Each week he has voluntary deductions of $8.80 for health care, $2.00 for life insurance, and $30.00 for credit union savings. What is his weekly net pay after income taxes, FICA, and voluntary deductions?

   _____

Using a Checking Account

Checks are a convenient and safe way to pay for purchases. You can use a checking account instead of cash to pay for goods and services. Many people don't like to carry large amounts of cash and pay for nearly everything with checks. Checks are safe because you can cancel them, or stop payment on them, if necessary. To keep track of your checking account, you'll need to use good math skills.

Making a Deposit

A **deposit** is money, either cash or checks, that you put in your account. When you open a new account, the bank assigns you an account number. To make a deposit, you must fill out a deposit ticket. A **deposit ticket**, or a deposit slip, provides your account number and lists the details of your transaction. Look at the sample that follows.

DEPOSITED IN		DOLLARS	CENTS
City Bank New York, New York 10001	CURRENCY	$ 300	00
	COINS	23	75
	1 List Checks Singly 63-689	156	87
DATE ___Aug. 1,___ 19 — —	2 31-269	19	90
Checks and other items are received for deposit subject to the terms and conditions of this bank's collection agreement.	3		
	4		
	5		
	Total From Other Side		
NAME ___Ellen K. Fitzpatrick___	SUBTOTAL	500	52
ACCOUNT NO. 430891-2	Less Cash Received	25	00
⑈0210⋯0005⑆	Net Deposit	$ 475	52

BE SURE EACH ITEM IS ENDORSED

EXAMPLE   On August 1, Ellen Fitzpatrick opened her account with a deposit of $300.00 in paper money, called **currency**, and $23.75 in coins. She also deposited two checks in the amounts of $156.87 and $19.90. Ellen would like to receive $25.00 in cash. Ellen's new account number is 430891-2.

Paper money and coins are shown on the Currency and Coin lines. The checks are listed separately on the Checks lines. The American Banking Association (ABA) number from each check being deposited is listed on the line to the left of each check amount. The ABA number identifies the location of the bank from which funds are being withdrawn. The ABA number is found on the upper right corner of checks near the date.

If Ellen had many checks to deposit, she could list them in the space provided on the other side of the deposit ticket. The total amount of the checks listed on the other side of the deposit ticket would be shown on the Total From Other Side line.

The Subtotal line shows the total of all money that is being deposited. The line Less Cash Received is filled in when a depositor wants to receive some cash back from the deposit. The Net Deposit line is a total of all currency, coins, and checks, less the amount of cash returned. It shows the total amount that is actually being deposited in the account.

Self-Check

Directions: Fill out the deposit slip that follows. You will deposit currency of $120.00 and coins of $5.50. Checks to be deposited are Check No. 42-220 for $49.50, Check No. 21-874 for $213.78, and ABA No. 21-356 for $49.33. Use your own name and today's date. Your Account No. is 536-652-9. Then compare your answer with that in the back of the book.

DEPOSITED IN				DOLLARS	CENTS
City Bank New York, New York 10001		CURRENCY			
		COINS			
		1 List Checks Singly			
DATE _____ 19 _____		2			
Checks and other items are received for deposit subject to the terms and conditions of this bank's collection agreement.		3			
		4			
		5			
NAME _____		Total From Other Side			
		SUBTOTAL			
ACCOUNT NO.	[ ]	Less Cash Received			
		Net Deposit			
⑈0210⬝⬝⬝0005⑈					

BE SURE EACH ITEM IS ENDORSED

Practice

Directions: Find the total amount deposited in each transaction. Then compare your answers with those in the back of the book.

	Currency Deposited	Coins Deposited	Checks Deposited	Cash Received	Total Amount Deposited
1.	$10.00	$5.00	$385.91	–	_____
2.	$77.00	$10.00	$967.33	–	_____
3.	$50.00	–	$104.00	$20.00	_____
4.	$65.00	$25.00	$610.56	$110.00	_____
5.	–	–	$934.21	$125.00	_____

Section 2

Maintaining Your Checkbook

After you make the first deposit in your checking account, you can write checks. Your **check** tells the bank to deduct money from your account to make a payment. Any time you write a check or make a deposit, keep track of these transactions in your **check register**.

Writing a Check

When you write a check, write clearly and fill out each space. Do not use pencil or someone may alter your check. In the example, Ellen Fitzpatrick is purchasing a gift at Stafford's Department Store.

1. Check number. Checks are numbered in a series for easy identification.

2. Date line. Always use the current date.

3. ABA identification number. The ABA number identifies the location of the bank from which funds are being withdrawn.

4. Payee. Write the name of the payee, the person or organization to whom payment will be made on the line beside *Pay to the Order of.* The payee receives the amount of money written in figures after the dollar sign on the same line.

5. Amount line. Write in words the amount of the check that appears in figures on the previous line with cents expressed as a fraction of a dollar. Draw a line that extends to the word, *Dollars.*

6. Signature line. Sign your name.

7. Memo line. Write the purpose of the check for your own records.

8. Bank number and account number. Number assigned by the bank.

Self-Check

Directions: Fill out the blank check on the next page with the information given. Make check number 110 payable to Welch Auto Repair in the amount of $111. The check is for fender repairs. Sign your own name and use the current date. Then compare your answer with that in the back of the book.

                                                                    NO. 110

                                        _____ 19 ____   1-5
                                                                       210

PAY TO THE
 ORDER OF _____ $_____

_____ DOLLARS

Memo: _____   _____

⑈0210⑈0005⑈   430742   3

Keeping Track of Checks

When a check is written or a deposit is made, a record should be kept either in a check register or on a check stub. A **check register** is a booklet in which you list checks written and deposits made. A **check stub** is the perforated portion of some checks. After you write a check or make a deposit, the new balance should be entered in your register or on your stub. If your register or stub is not up to date, you may overdraw the account. **Overdraw** means that you wrote a check for more money than you had in your account. When you overdraw your account, you create an overdraft. An **overdraft** is a check written for more than your account balance. Many banks charge a fee of $10 to $20 for each overdraft. Frequent overdrafts may affect your credit rating.

EXAMPLE

Lena Davis had a balance of $241.86 in her account. She wrote two checks, Check No. 86 for $74.87 on April 2 and Check No. 87 for $23.50 on April 5. Checks are deducted from the account balance. Lena deposited $125.60 on April 6. Deposits are added to the account balance. The register that follows records those transactions and gives her ending balance.

Number	Date	Description of Transaction	Payment Debit (−)	✓ T	Fee (If Any) (−)	Deposit Credit (+)	BALANCE	
							$ 241	86
86	4/2	Pat's Furniture	74 87				74	87
		new chair					166	99
87	4/5	Spray Shop	23 50				23	50
		pool supplies					143	49
	4/6	Deposit				125 60	125	60
							269	09

Self-Check

Directions: Three checks have been entered in a check register. Calculate the new balance after each check. Then compare your answer with that in the back of the book.

Number	Date	Description of Transaction	Payment Debit (−)	✓ T	Fee (If Any) (−)	Deposit Credit (+)	BALANCE $789 60
908	5/5	United Way charity	$25 01				
909	5/5	Mick Tiell yard work	$41 35				
910	5/5	Breezeway Apts. rent	$200 00				

Practice

Directions: Using the information that follows, write the transactions and calculate the ending balance in the check register. Then compare your answer with that in the back of the book. Beginning balance is $467.90.

6/13   Wrote Check No. 100 for $45.67 to Sara the Florist for plants.

6/15   Wrote Check No. 101 for $54.90 to Bill Jones for carpentry service.

6/20   Deposited $176.00 into account.

6/21   Wrote Check No. 102 for $45.60 to Crossman's for sneakers.

6/22   Wrote Check No. 103 for $10.00 to Colonial Theater for tickets and Check No. 104 for $15.00 to Washington Cleaners for dry cleaning.

Number	Date	Description of Transaction	Payment Debit (−)	✓ T	Fee (If Any) (−)	Deposit Credit (+)	BALANCE

Help with *Keeping an Accurate Check Register*

Here are some tips to help you keep an accurate check register:

1. Enter each check or deposit in your check register before you write the check or make the deposit so you don't forget to enter it later.

2. Subtract the check or add the deposit in your check register immediately. That way you will always know how much money is in your checking account.

3. Always date your check and the check register entry with the same date — the date on which you write the check. Some people think that a check written on a Sunday is not valid so they change the date. A check written on a Sunday is just as valid as one that is written any other day of the week. The date on the check should always match the date in the check register.

4. Always fill out the entire check. Never leave the payee's name or the amount blank.

5. If you make an error on a check, write "void" in ink across the front of the check. Make a corresponding entry in your check register, such as "#121, void." That way, you won't wonder why you have a check missing in your sequence of canceled checks.

6. If you use a cash machine for withdrawals, enter the amount of the withdrawal in your check register before you complete the transaction. Keep all cash machine receipts with your check register to help reconcile your bank statement.

Section 3

Reconciling Your Bank Statement

You will receive a statement from your bank on a regular basis. The **bank statement** summarizes your transactions for the previous statement period. The statement lists all transactions that the bank has received *until* the date of the statement. **Outstanding checks** and **deposits** are transactions that appear in your register but did not reach the bank in time to be processed and listed on your statement.

The bank statement also shows any service fees subtracted from your balance and any interest earned added to your balance. Many banks send canceled checks with the statement. **Canceled checks** are those that have been cashed. The balance in your register and the balance in the statement must agree. You should **reconcile**, or balance, the statement to make sure that it agrees with your check register each statement period to avoid overdrafts.

Linda Krosky uses her bank statement and the bank's *reconciliation form* on the back to balance her account. Here is what Linda's reconciliation looks like.

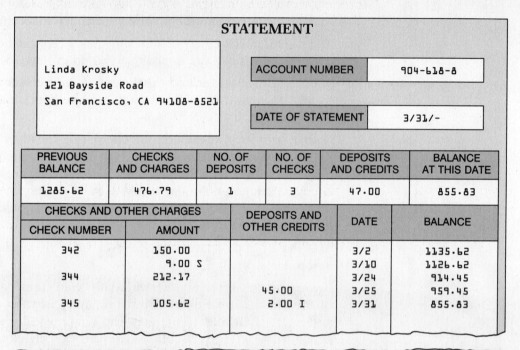

STATEMENT

Linda Krosky
121 Bayside Road
San Francisco, CA 94108-8521

ACCOUNT NUMBER	904-618-8

DATE OF STATEMENT	3/31/-

PREVIOUS BALANCE	CHECKS AND CHARGES	NO. OF DEPOSITS	NO. OF CHECKS	DEPOSITS AND CREDITS	BALANCE AT THIS DATE
1285.62	476.79	1	3	47.00	855.83

CHECKS AND OTHER CHARGES		DEPOSITS AND OTHER CREDITS	DATE	BALANCE
CHECK NUMBER	AMOUNT			
342	150.00		3/2	1135.62
	9.00 S		3/10	1126.62
344	212.17		3/24	914.45
		45.00	3/25	959.45
345	105.62	2.00 I	3/31	855.83

CHECKS OUTSTANDING NOT CHARGED TO ACCOUNT		
# 343	$ 102	65
346	43	62
347	87	96
TOTAL	234	23

ENDING BALANCE SHOWN	$ 855	83
ADD (+) DEPOSITS NOT YET CREDITED (IF ANY)	200	00
TOTAL	$ 1,055	83
SUBTRACT (−) CHECKS OUTSTANDING	234	23
THIS BALANCE SHOULD AGREE WITH YOUR CHECKBOOK BALANCE. BE SURE TO DEDUCT HANDLING COSTS, IF ANY, AND ADD INTEREST EARNED, IF ANY, TO CHECKBOOK BALANCE. BALANCE	$ 821	60

Her bank statement balance is $855.83. There are three outstanding checks: No. 343 for $102.65, No. 346 for $43.62, and No. 347 for $87.96. The outstanding checks are totaled and deducted from the bank statement balance. Linda has one outstanding deposit of $200.00. It is added to the bank balance. The adjusted balance of the bank statement is $821.60.

Bank Statement Balance − Outstanding Checks + Outstanding Deposits
= Adjusted Bank Statement Balance

The check register balance is $828.60. Earned interest of $2.00 is added to her check register balance and the service charge of $9.00 is deducted. Her new adjusted check register balance is $821.60, the same as her adjusted bank statement balance. She has reconciled her account.

Check Register Balance + Interest − Service Charges
= Adjusted Check Register Balance

Self-Check

Directions: Use the information given in each problem to determine the new check register balances and the adjusted bank statement balances. All the balances may not reconcile. Then check your answers with those in the back of the book.

	Check Register Balance	Bank Service Charge	New Register Balance	Bank Statement Balance	Out-standing Checks	Out-standing Deposits	Adjusted Statement Balance
1.	$823.55	$2.30	_____	$825.85	–	–	_____
2.	$424.47	$1.00	_____	$613.48	$190.01	–	_____
3.	$340.81	–	_____	$403.09	$587.76	$525.48	_____
4.	$950.94	$4.47	_____	$904.36	$257.93	$300.04	_____
5.	$90.06	$30.00	_____	$201.10	$191.04	$50.00	_____

Practice

Directions: Solve the problems that follow. Then compare your answers with those in the back of the book.

1. David Walker is reconciling his checking account. His bank statement balance is $989.27. He has an outstanding check of $49.77. His check register balance is $935.65. He earned interest of $3.85. What is his new check register balance? _____ What is his adjusted statement balance? _____

2. Joel Billingsley usually has $50 left in his checking account at the end of each month after paying his bills. His average daily balance is about $600 each month. Joel writes an average of nine checks per month. Joel can choose among three banks for a checking account. How much will he pay in a year's time for a checking account at each bank?
   (a) World Bank charges $.10 per check and has a $5.00 monthly service charge. _____
   (b) Bayside Bank has a flat service charge of $7.00 per month. _____
   (c) City Bank charges $.25 per check and a $2.50 monthly service charge. _____

3. Jim Phen is reconciling his checking account. His check register balance is $565.92. His bank statement balance shows a balance of $302.42. Interest of $2.15 has been earned. A fee of $15.63 for an overdraft has been charged. A deposit of $307.00 is outstanding. A check of $56.98 is outstanding. What is Jim's adjusted check register balance? _____ Bank statement balance? _____

Using a Calculator: *Balancing Your Checkbook*

Using the memory functions of your calculator can make the process of reconciling your bank statement balance with your checkbook register balance much easier.

Look at the following illustration which shows a portion of a checkbook register and a portion of a bank statement.

Number	Date	Description of Transaction	Payment Debit (−)	✓ T	Fee (If Any) (−)	Deposit Credit (+)	BALANCE	
							915	39
203	2/25	Fred Watson	29 90				29	90
		Carpool					885	49
204	2/21	Maxwell Gardens	15 79				15	79
		plants					869	70
205	2/27	Copy Store	49 00				49	00
		resume					820	70
	2/27	Deposit				105 00	105	00
							925	70

PREVIOUS BALANCE	CHECKS AND CHARGES	NO. OF DEPOSITS	NO. OF CHECKS	DEPOSITS AND CREDITS	BALANCE AT THIS DATE
970.69	55.30	0	2	4.50 I	919.89

CHECKS AND OTHER CHARGES		DEPOSITS AND OTHER CREDITS	DATE	BALANCE
CHECK NUMBER	AMOUNT			
			2/23	915.39
		4.50 I	2/28	919.89

Start with the check register balance. Press **9 2 5 . 7 0 M+** to enter this amount into memory.

Now, enter the bank statement balance. Press **9 1 9 . 8 9.** Then subtract the outstanding checks and add the unrecorded deposit. Press **− 2 9 . 9 0 − 1 5 . 7 9 − 4 9 . 0 0 + 1 0 5 . 0 0 − MR** (RM or memory recall) **=**. This string of calculations has adjusted and totaled the bank statement balance and subtracted it from the check register balance stored in memory. If the total shown is "0," the balances would be equal. If a positive balance is shown, the bank statement shows more money in the account. If a negative balance is shown, the check register shows more money in the account. In this case, the balance is positive, 4.50. The checkbook balance is off by that much. The first step in finding the difference is to look for any items equal to that amount. Interest in the amount of $4.50 was paid by the bank, but not entered into the checkbook register. By adding $4.50 to the checkbook register balance, the account will balance.

Unit 6 Review

Developing Your Skills

Directions: Follow the instructions for each set of problems.

Using the information given, calculate the total deposit for each problem.

	Currency	Coin	Checks	Cash Received	Total Deposit
1.	—	$7.00	$116.00	$20.00	_____
2.	$35.00	$5.00	$50.00	—	_____
3.	$100.00	—	$25.87, $589.87	—	_____
4.	—	—	$56.99, $29.99	$40.00	_____
5.	$50.00	—	$39.99, $17.56	$20.00	_____
6.	$20.00	$8.00	$59.99, $246.21	—	_____
7.	$246.50	—	$849.50, $892.10	$50.00	_____
8.	$50.00	$4.50	—	—	_____
9.	$20.00	—	$97.80, $44.50	$10.00	_____
10.	$39.00	$5.00	$99.41, $28.79	$10.00	_____

Using the information that follows, calculate each day's balance in the check register.

	Date	Deposits Made	Checks Written	Balance
				$321.07
11.	3/1	$1,704.21	—	_____
12.	3/7	—	$321.11, $110.06	_____
13.	3/11	—	$841.36	_____
14.	3/17	—	$18.21	_____
15.	3/21	$156.10	—	_____
16.	3/23	—	$77.12, $13.41	_____
17.	3/25	—	$20.00	_____
18.	3/25	—	$24.99	_____
19.	3/28	—	$33.98, $23.99	_____
20.	3/30	—	$14.31	_____

Use the information given in each problem to determine the new check register balances and the adjusted bank statement balances. The accounts may not reconcile.

	Check Register Balance	Interest	Bank Service Charge	New Register Balance	Bank Statement Balance	Out-standing Checks	Out-standing Deposits	Adjusted Statement Balance
21.	$209.33	—	$7.12	_____	$291.87	$89.66	—	_____
22.	$1,316.95	—	$17.44	_____	$99.10	—	$1,200.41	_____
23.	$604.30	—	—	_____	$648.50	$44.20	—	_____
24.	$1,304.40	—	$5.50	_____	$798.40	$49.50	$550.00	_____
25.	$310.93	—	$14.12	_____	$210.11	$12.40	$99.10	_____
26.	$2,133.25	—	$29.15	_____	$40.31	$36.21	$2,100.00	_____
27.	$1,178.81	$2.85	—	_____	$1,203.10	$21.44	—	_____
28.	$2,030.10	$4.10	$6.00	_____	$1,503.50	$21.30	$546.00	_____
29.	$737.77	$7.83	—	_____	$1,958.00	$1,212.00	—	_____
30.	$3,354.65	$3.80	—	_____	$1,249.87	$78.32	$2,186.90	_____

Solving Word Problems.

Directions: In solving a mathematics word problem, ask: (1) What information is being asked for? (2) What information is given in the problem? (3) What information is needed to solve the problem? Use the formulas in this unit to solve these problems.

31. Jason Clark deposited $50.00 in currency, $10.00 in coin, and a check for $298.51. How much did he deposit? _____

32. Claudette Kim made a deposit at the bank and later entered the deposit from memory in her checkbook register. She deposited $500.00 in currency; two checks for $175.85 and $14.15; and took cash back of $10.00. She thought she had deposited $645.00. What amount did she deposit? _____

33. Matthew Vincenti had a check register balance of $321.40. He deposited three checks in the amounts of $29.99, $421.00, and $33.41. He took cash back of $35.00. What is his new check register balance? _____

34. Marcella Wilson's bank statement showed a balance of $267.67. She had three outstanding checks: $42.11, $213.68, and $12.98. In addition, there was an outstanding deposit of $300.00 and a bank service charge of $7.00. When Marcella reconciled her check register, she had a balance of $305.90. Which item has Marcella forgotten to enter in her check register? _____

UNIT 7

Figuring Interest on Savings and Loans

When you deposit money in a savings account, the bank pays you a fee for the use of your money. Similarly, when you borrow money from a bank, you pay the bank a fee for the use of its money. Knowing how these fees are computed can help you shop around to get the most for the money you lend and pay the least for the money you borrow.

Figuring Simple Interest

In a sense, all the money you have on deposit in a savings account might be considered as "on loan" to the bank. The bank will use your and other depositors' money for a variety of purposes, including lending it out to others. For the use of your money, the bank pays you a fee. This fee, called **interest** (I), is based on the following three factors:

The *principal* (P), which is the amount of money deposited in the account.

The *rate* (R), which is the rate of interest you are paid.

The *time* (T), which is the length of time the money is deposited.

Interest paid on the original principal only is called **simple interest**. It is computed using this formula:

$$I = P \times R \times T$$

EXAMPLE

Jenny Carruthers put $500 into her savings account for 1 year. At $5\frac{3}{4}\%$ simple interest, how much has she earned at the end of the year?

$$I = P\ (\$500) \times R\ (5\tfrac{3}{4}\%) \times T\ (1\ year)$$
$$I = \$500 \quad\ \times .0575 \quad \times 1$$
$$I = \$28.75$$

Self-Check

Directions: Use the simple interest formula to solve the following problems. Round answers to the nearest cent. Then compare your answers with those in the back of the book.

1. Sanya Tarro has $3,573 in a savings account that earns $5\frac{3}{4}\%$ interest. After 1 year, how much simple interest did Sanya earn?

   _____

2. Maxwell Johnson has $100,000 in a savings account that earns $8\frac{1}{2}\%$ interest. After 1 year, how much did Maxwell earn in simple interest?

   _____

3. Ted Crisp has $2,634.09 in a savings account that earns $7\frac{1}{2}\%$ interest. After 1 year, how much did Ted earn in simple interest?

   _____

4. Jackson Giles has $10,521.89 in a savings account that earns $9\frac{1}{4}\%$ interest. After 1 year, how much simple interest did Jackson earn?

   _____

Interest Earned Stated in Months

In the previous examples, the amount of time was one year. Often the amount of time may be less than one year and stated in months. In this event, set up the time as a fraction, with the number of months of interest as the numerator. Because there are 12 months in the year, use 12 as the denominator. When it is convenient to do so, you may also convert the fraction to a decimal.

EXAMPLE

Ray Wobard has $135 on deposit in the Washington Cooperative Bank. His money is earning $5\frac{1}{4}$% interest. The interest is calculated 3 months later. How much interest has he earned?

$$I = P \text{ (\$135)} \times R \text{ } (5\tfrac{1}{4}\%) \times T \text{ } (\tfrac{3}{12} \text{ or } \tfrac{1}{4} \text{ year or .25})$$

$$I = \$135 \quad\quad \times .0525 \quad\quad \times \tfrac{1}{4} \text{ (or .25)}$$

$$I = \$1.77$$

Self-Check

Directions: Use the simple interest formula to solve the following problems. Round answers to the nearest cent. Then compare your answers with those in the back of the book.

1. Isaac Kimball has been collecting interest at $9\frac{3}{4}$% for the last 6 months on a $1,500 principal. How much interest has he earned?

   _____

2. Josie Rice invested $10,000 in an account that earns $8\frac{1}{2}$% interest. After 3 months, how much interest will Josie earn? _____

3. Carlton Armando has $14,989.56 in a savings account that earns 10% interest. After 9 months, how much simple interest will Carlton earn? _____

Solving for Other Parts of the Formula

The simple interest formula can be rearranged to find any missing part in the formula. Remember, when any two parts of a formula are known, you can find the unknown part. For example:

$$P = \frac{I}{R \times T} \quad\quad R = \frac{I}{P \times T} \quad\quad T = \frac{I}{P \times R}$$

EXAMPLE

What is the principal if the interest is $16.50, the interest rate is 6%, and the time is 2 months?

$$P = \frac{I}{R \times T}$$

$$P = \frac{\$16.50}{.06 \times \frac{2}{12}}$$

$$P = \frac{\$16.50}{.01}$$

$$P = \$1,650.00$$

Figuring Simple Interest **101**

EXAMPLE   If the principal is known to be $1,650 and the time to be 2 months, the rate would be determined as follows:

$$R = \frac{I}{P \times T}$$

$$R = \frac{\$16.50}{\$1,650 \times \frac{2}{12}}$$

$$R = .06, \text{ or } 6\%$$

EXAMPLE   Similarly, if the principal ($1,650) and the rate (6%) were known, the time would be calculated in this way:

$$T = \frac{I}{P \times R}$$

$$T = \frac{\$16.50}{\$1,650 \times .06}$$

$$T = .1667 \text{ of a year, or } \frac{2}{12}, \text{ or 2 months}$$

Note that the decimal answer here must be converted to months.

Self-Check

Directions: Use the simple interest formula to solve the following problems. Round answers to the nearest cent. Then compare your answers with those in the back of the book.

1. Ray Ming earned $15 in interest at a rate of 5% for 1 year. What is the amount of his principal? _____

2. Sally Lemoine earned $40 in interest on a principal of $250 in 2 years. What is the rate of interest? _____

3. Fred Turnkey invested a principal of $2,000 at an interest rate of 5% and earned $200 in interest. What length of time was he earning 5% interest? _____

Practice

Directions: Use the simple interest formula to solve the following problems. Round answers to the nearest cent. Then compare your answers with those in the back of the book.

1. Qon Yee has $1,750 in his savings account earning $5\frac{1}{2}\%$ interest. After 6 months, what amount of interest will he have earned?

_____

2. Oscar Heublein invested $1,430.75 in a savings account. His account paid $6\frac{3}{4}\%$ interest. What was the interest earned on Oscar's account after 1 year? _____

3. Cari Domingo has earned interest of $11 on her savings deposited at 6% interest for 3 years. What is the amount of principal?

_____

4. Philip Armbruster has $600 in his savings account earning $5\frac{3}{4}$% annual interest. After 12 months, what amount of interest will he have earned? _____

5. Brenda Carr put $1,500 into a savings account for 1 year and earned $120 in interest. What interest rate did she earn? _____

Using a Calculator: *The Memory Keys*

Memory is a calculator function that enables you to store and recall numbers. Memory is especially useful if you need to use the same number over and over again in calculations. The memory keys on a calculator are usually shown as M+, M−, MR or RM (memory recall), and MC or CM (memory clear). Check your calculator to determine what the memory keys are.

Suppose you want to calculate $6\frac{1}{2}$% simple interest for 1 year for these amounts: $500, $765, $3,200, $4,625, and $7,500. You want to know the total of the principal plus interest for each amount. Using your calculator, follow these steps:

Step 1	Press **1 . 0 6 5**	The 1 stands for the principal amount. The .065 is the interest that is added to the principal.
Step 2	Press **M+**	This puts 1.065 into memory.
Step 3	Press **5 0 0**	This is the first amount you want to calculate the interest for.
Step 4	Press **×**	This is to multiply $500 by the number in memory.
Step 5	Press **MR**	This recalls 1.065 from memory.
Step 6	Press **=**	This calculates the total of principal plus interest: $532.50.

To continue the calculations, repeat steps 3 through 6, entering the next amount. Press **7 6 5 × MR =**. The total of the principal plus interest is $814.73. Try the remaining amounts on your own. The answers are printed below.

The MR key can also be used to keep a running total. Add several numbers. Press **M+**. You will get a subtotal of the numbers that you have entered. You can keep adding numbers and get a subtotal of them any time by pressing **M+**. If you want to total all the numbers, press **MR**. What if you enter the wrong number? Use the **M−** key to subtract the entry from memory. However, every calculator is different. Work with your calculator to determine what special function keys you have and what they can do.

Answers $3,200.00 = $3,408.00; $4,625.00 = $4,925.63; $7,500.00 = $7,987.50

Figuring Compound Interest

Unlike simple interest, **compound interest** is interest paid on both the principal *and* the interest that has been paid in previous interest periods. Banks use different ways to calculate the interest they pay on savings. Interest paid on savings can be calculated daily, monthly, quarterly (once every three months), semiannually, or annually.

For example, you put $100 into a savings account that pays 6% interest annually. At the end of 1 year, you will have earned $6. If interest on your savings account is compounded semiannually, however, you will have earned $3 in interest at the end of 6 months. At the end of 1 year, you will have earned interest on the $100 principal and the $3 of interest that you earned at the end of 6 months. Here are the calculations:

$$I = P \times R \times T$$

First 6 Months $I = \$100.00 \times .06 \times \frac{1}{2}$

$I = \$3.00$

Second 6 Months $I = \$103.00 \times .06 \times \frac{1}{2}$

(Remember that now the principal has the interest for the previous period added on. The principal for the second 6 months, therefore, is $100.00 + $3.00, or $103.00.)

$I = \$3.09$

Interest Earned
After 1 Year $I = \$3.00 + \$3.09 = \$6.09$

Compound interest is often referred to as "interest on interest." Now let's see how much interest would be paid on the same amount if interest is compounded quarterly (every 3 months).

First 3 Months $I = \$100.00 \times .06 \times \frac{1}{4} = \$1.50$

Second 3 Months $I = \$101.50 \times .06 \times \frac{1}{4} = \$1.52$

Third 3 Months $I = \$103.02 \times .06 \times \frac{1}{4} = \$1.55$

Fourth 3 Months $I = \$104.57 \times .06 \times \frac{1}{4} = \$1.57$

Interest Earned
After 1 Year $I = \$1.50 + \$1.52 + \$1.55 + \$1.57 = \$6.14$

Using a Compound Interest Table

What you have just done in the previous example is to compute compound interest the hard way! A much easier way is to use a compound interest table such as the one that follows.

AMOUNT OF $1.00							
TOTAL INTEREST PERIODS	INTEREST RATE PER PERIOD						
	1.2500%	1.3750%	1.5000%	2.7500%	2.8750%	3.0000%	3.1250%
1	1.01250	1.01375	1.01500	1.02749	1.02875	1.03000	1.03125
2	1.02515	1.02768	1.03022	1.05575	1.05832	1.06090	1.06347
3	1.03797	1.04182	1.04567	1.08478	1.08875	1.09272	1.09671
4	1.05094	1.05614	1.06136	1.11462	1.12005	1.12550	1.13098
5	1.06408	1.07066	1.07728	1.14527	1.15225	1.15927	1.16632
6	1.07738	1.08538	1.09344	1.17676	1.18538	1.19405	1.20277
7	1.09085	1.10031	1.10984	1.20912	1.21946	1.22987	1.24036
8	1.10448	1.11544	1.12649	1.24237	1.25452	1.26677	1.27912
9	1.11829	1.13078	1.14339	1.27654	1.29059	1.30477	1.31909
10	1.13227	1.14632	1.16054	1.31165	1.32769	1.34391	1.36031
11	1.14642	1.16209	1.17795	1.34772	1.36586	1.38423	1.40282
12	1.16075	1.17806	1.19562	1.38478	1.40513	1.42576	1.44666
13	1.17526	1.19426	1.21355	1.42286	1.44553	1.46853	1.49187
14	1.18995	1.21068	1.23175	1.46199	1.48709	1.51258	1.53849
15	1.20482	1.22733	1.25023	1.50219	1.52984	1.55796	1.58657
16	1.21988	1.24421	1.26898	1.54350	1.57382	1.60470	1.63615
17	1.23513	1.26132	1.28802	1.58595	1.61907	1.65284	1.68728
18	1.25057	1.27866	1.30734	1.62956	1.66562	1.70243	1.74000
19	1.26620	1.29624	1.32695	1.67438	1.71351	1.75350	1.79438
20	1.28203	1.31406	1.34685	1.72042	1.76277	1.80611	1.85045
21	1.29806	1.33213	1.36706	1.76773	1.81345	1.86029	1.90828
22	1.31428	1.35045	1.38756	1.81635	1.86559	1.91610	1.96791
23	1.33071	1.36902	1.40838	1.86630	1.91922	1.97358	2.02941
24	1.34735	1.38784	1.42950	1.91762	1.97440	2.03279	2.09283

Let's look at an example. Take the same amount of $100 at 6% interest, compounded quarterly for 1 year.

Step 1   Find the total number of interest periods in a year.

12 months ÷ 3 = 4 quarters = 4 interest periods

Step 2   Find the interest rate per period.

6% ÷ 4 periods = 1.5% interest rate per period

Step 3   Find the interest amount for $1.00 from the table.

In the column "Total Interest Period," read down until you locate the number 4. Then read across that line to the "1.5000%" column. The answer is 1.06136.

Step 4   Find the amount.

$100 × 1.06136 = $106.136 (Round to $106.14.)

The amount of interest earned on $100.00 in 1 year at 6%, compounded quarterly, is $6.14.

Self-Check

Directions: Use the interest table to solve the following problems. Then compare your answers with those in the back of the book.

1. Salvatore Johnson has $500 in a savings account that earns $5\frac{1}{2}$% interest, compounded quarterly. After 1 year, how much interest has Salvatore earned? _____

2. Walter Horowitz invested $300 in a fund collecting 11% interest compounded quarterly. After 1 year, how much compound interest has Walter earned? _____

3. Mercedes Robiero has $2,000 in a savings account that collects 12.5% interest compounded quarterly. After 1 year, how much interest has Mercedes earned? _____

4. Samantha Bellingham has $2,600 in a fund collecting 11% interest compounded quarterly. After 1 year, how much compound interest has Samantha earned? _____

5. Al Bemelmans has $15,000 in a savings account collecting 12% interest compounded quarterly. After 1 year, how much compound interest has Al earned? _____

Interest Compounded for More Than a Year

Let's take the amount of $100 again. As before, let's use an interest rate of 6%, compounded quarterly. However, this time let's use a time period of 3 years. Note there is an additional step—Step 3.

Step 1 Find the total number of interest periods in a year.

12 months ÷ 3 = 4 quarters = 4 interest periods

Step 2 Find the interest rate per period.

6% ÷ 4 periods = 1.5% interest rate per period

Step 3 Find the number of interest periods in all the years.

4 quarters × 3 years = 12 periods

Step 4 Find the interest amount for $1.00 from the table on page 105.

In the column "Total Interest Period," read down until you locate the number 12. Then read across that line to the "1.5000%" column. The answer is 1.19562.

Step 5 Find the amount.

$100.00 × 1.19562 = $119.562 (Round to $119.56.)

The amount of interest earned on $100.00 in 3 years at 6%, compounded quarterly, is $19.56.

Self-Check

Directions: Use the interest table on page 105 to solve the following problems. Then compare your answers with those in the back of the book.

1. Veronica Giles has $500 in a savings account earning 12% interest compounded quarterly. After 2 years, how much compound interest has Veronica earned? _____

2. Edward Buruca has $10,000 in a savings account earning 11% interest compounded quarterly. After 4 years, how much compound interest has Edward earned? _____

3. Patty Dudek deposited $25,260.49 in a savings account earning 12.5% interest compounded quarterly. After 3 years, how much compound interest would Patty earn? _____

Interest Compounded Daily

Sometimes money can be invested at an interest rate that is compounded daily. Each day the account balance is changed by adding the interest earned to the principal. Then that new balance is used in calculating the *next* day's interest. Use the daily compound interest table that follows to find the amount of interest earned in the example.

	AMOUNT OF $1.00 AT 5.5%, COMPOUNDED DAILY, 365-DAY YEAR								
DAY	**AMOUNT**	**DAY**	**AMOUNT**	**DAY**	**AMOUNT**	**DAY**	**AMOUNT**	**DAY**	**AMOUNT**
1	1.00015	11	1.00165	21	1.00316	31	1.00468	50	1.00755
2	1.00030	12	1.00180	22	1.00331	32	1.00483	60	1.00907
3	1.00045	13	1.00196	23	1.00347	33	1.00498	70	1.01059
4	1.00060	14	1.00211	24	1.00362	34	1.00513	80	1.01212
5	1.00075	15	1.00226	25	1.00377	35	1.00528	90	1.01364
6	1.00090	16	1.00241	26	1.00392	36	1.00543	100	1.01517
7	1.00105	17	1.00256	27	1.00407	37	1.00558	110	1.01670
8	1.00120	18	1.00271	28	1.00422	38	1.00574	120	1.01823
9	1.00135	19	1.00286	29	1.00437	39	1.00589	130	1.01977
10	1.00150	20	1.00301	30	1.00452	40	1.00604	140	1.02131

EXAMPLE

Andy McHenry put $1,125 into his savings account on June 2, intending to draw it out on August 31 to help pay for his college tuition. His savings account pays 5.5% interest compounded daily. What will be the amount in his savings account on August 31?

Step 1 Find the number of days.

From June 2 to August 31 = 90 days.
When calculating days, do not count the day of deposit.

Step 2 Use the table to find the interest amount for $1.00 at 5.5% compounded daily for 90 days: 1.01364.

Step 3 Multiply the principal by the interest amount for $1.00 from the table: $1,125.00 × 1.01364 = $1,140.345. Round to the nearest cent. The amount in Andy's savings account on August 31 will be $1,140.35.

He has earned $15.35 in interest.

Self-Check

Directions: Use the daily compound interest table on page 107 to solve the following problems. Then compare your answers with those in the back of the book.

1. Sam Gallegas had $5,350 in a savings account from March 2 to May 31. His account pays 5.5% interest compounded daily. How much did Sam have in his savings account after this time period?

   _____

2. Gail Rottman put $25,000 into a savings account, from October 1 through November 10, that earned 5.5% interest compounded daily. After this time period, how much interest did Gail earn?

   _____

3. Peter Lennon had $500 in a savings account from July 5 to August 10. He earned 5.5% interest compounded daily. How much did Peter have in his account after this time period? _____

4. Vincent Washington put $14,200 in a savings account from March 15 through May 4. His account earned 5.5% interest compounded daily. How much did Vincent have in his savings account after this time period? _____

Practice

Directions: Use the compound interest table on page 105 and daily compound interest table on page 107 to figure the amount of interest earned in the following problems. Then compare your answers with those in the back of the book.

1. Jane Sprague invested $3,000 in a fund that paid 12% interest, compounded quarterly. After 1 year, how much interest has she earned? _____

2. Marge Langone has had $2,750 in her savings account for 6 months. Her account pays 5.5% interest compounded quarterly. How much interest has she earned? _____

3. If Kelly DeForest can put $3,500 in an investment fund for 3 years, she can earn 11% interest compounded quarterly. How much interest will she earn? _____

4. Ten years ago, Sally De Vito invested $2,370 at $6\frac{1}{4}$% interest, compounded semiannually. How much interest has she earned during the 10 years? _____

5. Bill Lazarus put $3,000 into his savings account and withdrew it after 4 months (120 days). His account paid 5.5% interest, compounded daily. How much interest had he earned?

   _____

Figuring the Cost of a Loan

People and businesses often borrow money from a bank or other financial institutions. When money is borrowed, the borrower signs a written promise to repay the amount of the loan. Such a promise to pay is called a **promissory note**. The promissory note states the amount borrowed (the principal), the interest rate, and the period of time that the money will be borrowed.

Types of Interest

Banks often calculate interest on their loans in different ways. Some banks charge **ordinary interest**, which is based on a 360-day year. Others charge **exact interest**, which is based on a 365-day year. Let's look at some examples.

EXAMPLE

Oxford Company borrowed $5,000 for 60 days at 11% interest.

Exact Interest Method:

$I = P \times R \times T$

$I = \$5,000.00 \times .11 \times \frac{60}{365}$

$I = \$550.00 \times \frac{60}{365}$

$I = \$90.41$

Ordinary Interest Method:

$I = P \times R \times T$

$I = \$5,000.00 \times .11 \times \frac{60}{360}$

$I = \$550.00 \times \frac{60}{360}$ (or $\frac{1}{6}$)

$I = \$91.67$

As you can see, the exact interest method results in a slightly lower interest charge. The ordinary interest method is used because it is generally easier to calculate interest with that method. Today, however, most banks use computers to calculate interest charges, so the 365-day year is often used.

Self-Check

Directions: Solve the following problems. Then compare your answers with those in the back of the book.

1. Walter Romerez borrowed $1,500 for 90 days at 12% interest from National Bank. Use the ordinary interest method to compute the interest Walter will owe the bank. _____

2. Ginger LeRoyer borrowed $2,000 for 60 days at 11% interest from Western Bank. Use the exact interest method to compute the interest that Ginger will owe the bank. _____

3. The Marcomb Company borrowed $25,000 for 80 days at 10% interest from City Trust Bank. Use the ordinary interest method to compute how much interest will be paid to the bank.

_____

Single-Payment Loans

Many personal loans and some business loans are single-payment loans. As the name suggests, these loans are to be repaid in one payment at the due date. The **due date** is the time at which the loan must be repaid. The due date is determined by the **term** of the loan, which is the amount of time for which the money is borrowed.

When a single-payment loan is repaid, the amount repaid, called the **maturity value**, includes both the principal and the interest.

EXAMPLE

Jake Welby borrowed $2,000 from his bank to pay for seed needed on his farm. The loan was to be repaid in 3 months (90 days) when he harvested his vegetables. The bank charged $12\frac{1}{2}\%$ interest. Figure the maturity value of Jake Welby's loan using the ordinary interest method.

$$I = P \times R \times T$$
$$I = \$2,000.00 \times .125 \times \frac{90}{360}$$
$$I = \$62.50$$

$$Maturity\ Value = Principal + Interest$$
$$Maturity\ Value = \$2,000.00 + \$62.50$$
$$Maturity\ Value = \$2,062.50$$

Self-Check

Directions: Figure the interest and maturity value of the following loans. Use the exact interest method. Then compare your answers with those in the back of the book.

	Principal	Rate	Term	Interest	Maturity Value
1.	$1,600	8%	240 days	_____	_____
2.	$3,500	10%	90 days	_____	_____
3.	$869	11%	72 days	_____	_____
4.	$5,275	9%	55 days	_____	_____
5.	$2,700	$14\frac{1}{4}\%$	180 days	_____	_____

Discounted Loans

Sometimes a lender prefers that a borrower pay the interest on a loan up front. Such loans are called **discount loans**. Here is how a discount loan works.

The borrower signs a promissory note for the amount of the loan, called the **face value** of the note. The interest, which is called the **bank discount**, is then deducted from the face value. When expressed as a percent, the bank discount is called the **bank discount rate**. The amount of money remaining after the bank discount is called the **proceeds**. The proceeds are the amount the borrower actually receives. Such a loan or note is called **non-interest bearing** because no further interest is charged after the initial bank discount. To figure a bank discount on a non-interest-bearing note, use the basic formula for determining simple interest.

Interest (Bank Discount) = Principal × Bank Discount Rate × Time

EXAMPLE   Findlay Industries needs to borrow $13,000 to meet seasonal operating expenses. The bank discount rate is 10%. It is April 15, and Findlay wants to repay the loan by May 30. With ordinary interest, what are the proceeds of Findlay's loan?

Step 1   Find the term of the loan.

$$April\ 15\ to\ April\ 30 = 15\ days$$
$$May\ 1\ to\ May\ 30 = 30\ days$$
$$Total = 45\ days$$

Step 2   Find the bank discount.

$$I = P \times R \times T$$
$$I = \$13,000.00 \times 10\% \times \frac{45}{360}$$
$$I = \$162.50$$

Step 3   Find the proceeds.

Face Value − Bank Discount = Proceeds
$$\$13,000.00 - \$162.50 \qquad = \$12,837.50$$

Self-Check

Directions: Figure the bank discount and proceeds of the following discounted loans. Use the ordinary interest method. Then compare your answers with those in the back of the book.

	Face Value	Bank Discount Rate	Time	Bank Discount	Proceeds
1.	$1,840	12.5%	100 days	_____	_____
2.	$2,000	16%	60 days	_____	_____
3.	$4,500	13.25%	240 days	_____	_____
4.	$650	10%	30 days	_____	_____
5.	$7,000	14%	90 days	_____	_____

Practice

Directions: Solve the following problems. Then compare your answers with those in the back of the book.

1. Bedford Company borrowed $100,000 for 60 days at 11%. Use the exact interest method to figure the interest. _____

2. Susan Rubacky borrowed $4,000 from her bank to pay for her computer. The loan was to be repaid in 90 days. The bank charged 11% interest. Figure the maturity value of Susan's loan using the ordinary interest method. _____

3. The Lopez Company needs to borrow $18,000 to meet operating expenses. The bank discount rate is 12%. It is October 3, and the Lopez Company wants to repay the loan by December 15. With ordinary interest, what are the proceeds of the Lopez Company's loan? _____

4. Katrina Silva borrowed $500 for 60 days at 12% interest from her bank. Using the exact method, what is the interest Katrina will owe her bank? _____

5. Julio Wright borrowed $5,000 from his bank and repaid it in 100 days. His bank charged 9% exact interest. Figure the maturity value of Julio's loan. _____

Help with *Managing Your Money*

When shopping for a car loan, be sure you know how much you will actually pay for the car before you sign the agreement. For example, suppose you want to buy a car that costs $12,900. The interest rate for the car loan is 11.9%. You can finance the loan for 60 months (5 years), 48 months (4 years), or 36 months (3 years). How much will you really pay for the car if you finance the entire amount of $12,900?

Number of Years	Monthly Payment	Finance Charge	Total Price
5	$342.93	$7,675.50	$20,575.50
4	396.68	6,140.40	19,040.40
3	486.25	4,605.30	17,505.30

Just imagine—if you finance for 5 years, you will pay over $20,000 for a car that costs $12,900! Now, let's assume you make a down payment of $1,500. How will financing only $11,400 affect the cost of the car?

Number of Years	Monthly Payment	Finance Charge	Total Price
5	$303.05	$6,783.00	$18,183.00
4	350.55	5,426.40	16,826.40
3	429.72	4,069.80	15,469.80

With a down payment of $1,500.00, you can save $2,392.50 over 5 years. If you leave the $1,500.00 in a savings account that pays 5.5% interest, at the end of 5 years you would have $1,912.50 in your account. By using the money to buy the car, you would not only "earn" $480.00, but you would also reduce the monthly payment by almost $40.00.

Unit 7 Review

Developing Your Skills

Directions: Follow the instructions for each set of problems. Use the formulas for interest to solve these problems. In addition, you may have to use the tables on pages 105 and 107.

Fill in the blanks using the simple interest formulas. Use ordinary interest. Round answers.

	Principal	Rate	Time	Interest
1.	$575	7%	2 years	_____
2.	$4,390	6%	1 year	_____
3.	_____	6%	66 days	$16.50
4.	$2,500	_____	6 months	$150.00
5.	$1,000	12%	_____	$60.00

Calculate the amount of interest for each of the following problems using the compound interest table and the daily interest table on pages 105 and 107.

6. $3,500.00 at 6% for 3 years, interest compounded semiannually.

_____

7. $600.00 at 12% for 6 years, interest compounded quarterly.

_____

8. $325.00 at 5.5% for 100 days, interest compounded daily.

_____

Solve the following problems for single-payment loans. Use ordinary interest. Fill in the missing amounts in the spaces provided.

	Principal	Rate	Term	Interest	Maturity Value
9.	$400	8%	90 days	_____	_____
10.	$2,400	11%	45 days	_____	_____
11.	$1,560	12%	120 days	_____	_____
12.	$880	10.5%	144 days	_____	_____
13.	$650	13.25%	63 days	_____	_____

Solving Word Problems

Directions: In solving a mathematics word problem, ask: (1) What information is being asked for? (2) What information is given in the problem? (3) What information is needed to solve the problem? You may need to use the formulas and tables provided in this unit to help solve the problems.

14. Lavinia Davis put $2,356 into a savings account for a year. If the account pays 10% simple interest, how much has she earned at the end of the year? _____

15. Kate Ryker deposits $500 in a savings account that earns 8.75% simple interest. She withdraws the money after 6 months. How much interest has she earned? _____

16. Carlos Rodriguez deposited $1,000.00 in a savings account for 1 month. He earned $6.04 in interest. What was the rate of simple interest that Carlos earned? _____

17. Beverly Chillingsworth wants to borrow $10,000 for 180 days. First Bank charges 14% ordinary interest. State Bank charges 13% exact interest. How much will Beverly pay if she borrows from First Bank? _____
How much will Beverly pay if she borrows from State Bank?

_____

18. Sal Fuentes applies for a discounted loan of $3,500 at 10% for 54 days at exact interest. What are the proceeds? _____
Bank discount? _____

19. Emily LaMotta put $470 into her savings account on July 1 and withdrew it 90 days later. Her account pays $5\frac{1}{2}$% annual interest, compounded daily. How much interest did Emily earn?

_____

20. On January 1 Gregory Olcott opened a savings account with a deposit of $768.50 at 5.5% annual interest compounded daily. On March 2 he withdrew his savings. How much interest had he earned? _____

21. Sally Hermann's dress shop needed a short-term loan of $4,550 to pay for a shipment. The bank gave her 144 days to repay at an interest rate of $9\frac{3}{4}$%, ordinary interest. What is the maturity value of Sally's loan? _____

22. Carol Crow put $4,000 into her credit union for two years at 5.5% compounded quarterly. How much interest did she earn?

_____

Using Credit

Often our desire to immediately enjoy certain goods and services is not matched by our available cash to pay for them. In these instances, we may want to consider using credit. Credit allows people to buy a good or service now and pay for it later. Without credit, many of us would have to delay making major purchases. There are many different types of credit. Shop for credit just as you would for any other product.

115

Buying on Installment

A very common form of credit is an **installment loan**, a loan repaid in equal amounts, or installments, over a specified period of time. Installment loans are most often made by consumers for the purchase of automobiles, large appliances, and furniture. A **down payment**, which is a percentage or portion of the total price, is usually required. The balance after the down payment is the **amount financed**, the loan amount.

Annual Percentage Rate

The interest on an installment loan is called the **finance charge**. The finance charge is included in the installment payments. The amount of the finance charge depends on three things: (1) the amount financed, (2) the number of monthly installments, and (3) the annual percentage rate charged. The **annual percentage rate (APR)** gives the true interest rate in terms of a percentage for the use of credit per year.

Because APR calculations are complex, annual percentage rate tables, like the one shown here, are often used to find the APR of monthly installment payments.

APR	Term in Months	If You Finance. . . $200	$500	$1000	$1500
		Your **Monthly Payments** Are			
10%	6	34.31	85.78	171.56	257.34
	12	17.58	43.95	87.91	131.87
	18	12.01	30.02	60.05	90.08
	24	9.22	23.07	46.14	69.21
15%	6	34.80	87.01	174.03	261.05
	12	18.05	45.12	90.25	135.38
	18	12.47	31.19	62.38	93.57
	24	9.69	24.24	48.48	72.72
18%	6	35.10	87.76	175.52	263.28
	12	18.33	45.84	91.68	137.52
	18	12.76	31.90	63.80	95.70
	24	9.98	24.96	49.92	74.88

EXAMPLE

Let's say you make an installment loan of $1,000 to pay for furniture. You intend to repay the loan in 12 equal monthly installments. The APR is 15%. Look under the APR column of the APR table and find 15%. Find the term, 12 months. Move across the columns to the amount financed, $1,000. The monthly payment amount is $90.25. Figure the finance charge.

Monthly Payment × Number of Payments = Total Amount Repaid
$90.25 × 12 = $1,083.00

Total Amount Repaid − Amount Financed = Finance Charge
$1,083.00 − $1,000.00 = $83.00

The finance charge is $83.00.

To determine the APR when the amount financed and the finance charges are known, another formula and APR table can be used.

EXAMPLE   Chris McCloud bought a computer for $5,000. Her finance charge for the 24-month installment purchase was $991. What was the APR?

First, find the finance charge per $100 of the loan.

$$\$100 \times \frac{\text{Finance Charge}}{\text{Amount Financed}} = \text{Finance charge per } \$100$$

$$\$100 \times \frac{\$991}{\$5,000} = \$19.82$$

Refer to the table titled "Annual Percentage Rates." Under the column headed *Term*, find 24 and read across *all* of the "24" row until you come to either the number $19.82 or a figure very close to it. *Read up*. What is the APR at the top of the column? It is 18%, Chris McCloud's APR.

ANNUAL PERCENTAGE RATES										
APR	**10.00%**	**10.25%**	**10.50%**	**10.75%**	**11.00%**	**11.25%**	**11.50%**	**11.75%**	**12.00%**	**12.25%**	**12.50%**
TERM	FINANCE CHARGE PER $100 OF AMOUNT FINANCED										
6	2.94	3.01	3.08	3.16	3.23	3.31	3.38	3.45	3.53	3.60	3.68
12	5.50	5.64	5.78	5.92	6.06	6.20	6.34	6.48	6.62	6.76	6.90
18	8.10	8.31	8.52	8.73	8.93	9.14	9.35	9.56	9.77	9.98	10.19
24	10.75	11.02	11.30	11.58	11.86	12.14	12.42	12.70	12.98	13.26	13.54
30	13.43	13.78	14.13	14.48	14.83	15.19	15.54	15.89	16.24	16.60	16.95
36	16.16	16.58	17.01	17.43	17.86	18.29	18.71	19.14	19.57	20.00	20.43

APR	**12.75%**	**13.00%**	**13.25%**	**13.50%**	**13.75%**	**14.00%**	**14.25%**	**14.50%**	**14.75%**	**15.00%**	**15.25%**
6	3.75	3.83	3.90	3.97	4.05	4.12	4.20	4.27	4.35	4.42	4.49
12	7.04	7.18	7.32	7.46	7.60	7.74	7.89	8.03	8.17	8.31	8.45
18	10.40	10.61	10.82	11.03	11.24	11.45	11.66	11.87	12.08	12.29	12.50
24	13.82	14.10	14.38	14.66	14.95	15.23	15.51	15.80	16.08	16.37	16.65
30	17.31	17.66	18.02	18.38	18.74	19.10	19.45	19.81	20.17	20.54	20.90
36	20.86	21.30	21.73	22.17	22.60	23.04	23.48	23.92	24.35	24.80	25.24

APR	**15.50%**	**15.75%**	**16.00%**	**16.25%**	**16.50%**	**16.75%**	**17.00%**	**17.25%**	**17.50%**	**17.75%**	**18.00%**
6	4.57	4.64	4.72	4.79	4.87	4.94	5.02	5.09	5.17	5.24	5.32
12	8.59	8.74	8.88	9.02	9.16	9.30	9.45	9.59	9.73	9.87	10.02
18	12.72	12.93	13.14	13.35	13.57	13.78	13.99	14.21	14.42	14.64	14.85
24	16.94	17.22	17.51	17.80	18.09	18.37	18.66	18.95	19.24	19.53	19.82
30	21.26	21.62	21.99	22.35	22.72	23.08	23.45	23.81	24.18	24.55	24.92
36	25.68	26.12	26.57	27.01	27.46	27.90	28.35	28.80	29.25	29.70	30.15

APR	**18.25%**	**18.50%**	**18.75%**	**19.00%**	**19.25%**	**19.50%**	**19.75%**	**20.00%**	**20.25%**	**20.50%**	**20.75%**
6	5.39	5.46	5.54	5.61	5.69	5.76	5.84	5.91	5.99	6.06	6.14
12	10.16	10.30	10.44	10.59	10.73	10.87	11.02	11.16	11.31	11.45	11.59
18	15.06	15.28	15.49	15.71	15.93	16.14	16.36	16.57	16.79	17.01	17.22
24	20.11	20.40	20.69	20.98	21.27	21.56	21.86	22.15	22.44	22.74	23.03
30	25.29	25.66	26.03	26.40	26.77	27.14	27.52	27.89	28.26	28.64	29.01
36	30.60	31.05	31.51	31.96	32.42	32.87	33.33	33.79	34.25	34.71	35.17

APR	**21.00%**	**21.25%**	**21.50%**	**21.75%**	**22.00%**	**22.25%**	**22.50%**	**22.75%**	**23.00%**	**23.25%**	**23.50%**
6	6.21	6.29	6.36	6.44	6.51	6.59	6.66	6.74	6.81	6.89	6.96
12	11.74	11.88	12.02	12.17	12.31	12.46	12.60	12.75	12.89	13.04	13.18
18	17.44	17.66	17.88	18.09	18.31	18.53	18.75	18.97	19.19	19.41	19.62
24	23.33	23.62	23.92	24.21	24.51	24.80	25.10	25.40	25.70	25.99	26.29
30	29.39	29.77	30.14	30.52	30.90	31.28	31.66	32.04	32.42	32.80	33.18
36	35.63	36.09	36.56	37.02	37.49	37.95	38.42	38.89	39.36	39.82	40.29

Self-Check

Directions: Use the tables on pages 116 and 117 and the formulas that you have just learned to solve the following problems. Then compare your answers with those in the back of the book.

1. Heather Courtney bought a stereo set for $1,800. She made a down payment of $300 and agreed to repay the loan in 24 months at an APR of 15%. What is the finance charge? _____

2. Len Garrity is paying $500.00 for a power lawn mower. The store's installment plan requires 18 monthly payments of $31.19 each. What is the APR on Len's loan? _____

3. Suburban TV is offering VCRs for $399, 18 months to repay the loan, and "only" $61 in finance charges. What is the APR for these loans? _____

Practice

Directions: Use the tables on pages 116 and 117 and the formulas that you have just learned to solve the following problems. Then compare your answers with those in the back of the book.

1. The Wards, Peggy and Brent, are buying furniture for a total of $4,300. The furniture store offered them a 24-month installment loan with finance charges of $790. What is the APR? _____

2. Hal and Mary Wisnowski were shopping for the best APR for an installment loan of $3,600 to buy furniture. Reliance Loan Company offered a 15% loan to be repaid in 36 months. Northwestern Mutual Bank offered a 17.5% loan to be repaid in 30 months. What was the finance charge for the Reliance installment loan? _____

   For the Northwestern Mutual loan? _____

   What is the actual difference in finance charges? _____

3. Jim Salvatore bought a washer and dryer for $1,850. He made a down payment of $350 and agreed to pay the loan in 24 months at an APR of 18%. What is the finance charge? _____

4. Paul Criscione is purchasing an electric stove with an installment loan. The electric stove sells for $700.00. He agreed to make a down payment of $200.00 and to make 24 monthly payments of $24.24 each. What is the finance charge? _____

5. Tanya Balboa purchased a stereo system that costs $1,500. She made a down payment of $500 and agreed to repay a loan from the stereo store in 18 months at an APR of 18%. What is her finance charge? _____

Using Credit Cards and Charge Accounts

Many retail businesses accept multipurpose credit cards, such as VISA and MasterCard, which may be issued by banks and other financial institutions. Other businesses may only accept their own cards, which are issued to their charge-account customers for use only in their stores. In either instance, the procedures are fairly similar and usually involve sales receipts, monthly statements, and established lines of credit.

Sales Receipts

When a buyer makes a credit-card purchase, the seller gives the buyer a sales receipt, just as if the buyer had paid cash. The **sales receipt** shows the buyer's name, lists the items that were purchased and the price of each item, the sales tax, and the total purchase price. The buyer signs the sales receipt and is given a copy of it as a record of the purchase. Look at the receipt below.

K 7605 3821F					
Georgette Williamson					
LONGWOOD DEPARTMENT STORE					

DATE 9/24	AUTHORIZATION NO.	SALES CLERK EF	DEPT. 29	IDENTIFICATION	TAKE ✓ SEND
QUAN	CLASS	DESCRIPTION		PRICE	AMOUNT
1		dress		24.99	$24.99
2		pair of socks		3.50	$ 7.00
1		belt		4.42	$ 4.42
				SUB TOTAL	$ 36.41
				SALES TAX	$ 1.82
				TOTAL	$ 38.23

The issuer of the card identified on this item is authorized to pay the amount shown as TOTAL upon proper presentation. I promise to pay such TOTAL (together with any other charges due thereon) subject to and in accordance with the agreement governing the use of such card.

CARDHOLDER SIGN HERE X *Georgette Williamson*

Self-Check

Directions: During the month of January, Pablo moved into an apartment and purchased new items for his kitchen. Using the information given in the table and a 5% sales tax, figure the sales tax and total for each separate shopping day. Compare your answers with those in the back of the book.

	Date	Quantity	Item Purchased	Item Price	Sales Tax	Total
1.	1/3	1	set of glasses	$32.99	_____	_____
2.	1/6	2	frying pans	$15.00	_____	_____
3.	1/15	1	set of dishes	$49.00	_____	_____
4.	2/24	6	placemats	$ 3.75	_____	_____
5.	2/26	8	soup bowls	$ 4.25	_____	_____

At the end of a billing cycle—which usually is 30 or 31 days in length— businesses total the purchases for each credit customer. They then send each credit card customer a monthly statement. A **monthly statement** is a summary of transactions for the billing cycle. The statement shows the previous balance, payments, new purchases, credits for returned merchandise, minimum payment due, and finance charge on the previous balance. Usually there is no finance charge if the total amount is paid within 30 days. Look at the monthly statement for Georgette Williamson's charge account.

LONGWOOD DEPARTMENT STORE			AMOUNT DUE THIS STATEMENT	18.00
GEORGETTE WILLIAMSON 83 LONGWOOD DRIVE MIDDLETON, MA 01949			NEW BALANCE	359.20
			$ AMOUNT PAID	

YOUR LINE OF CREDIT $1,000   AVAILABLE CREDIT IS $640

Mo.	Day	Reference	TRANSACTION DESCRIPTION See reverse for detailed description of department numbers indicated below.	CHARGES	PAYMENTS & CREDITS
			**FINANCE CHARGE ON AVERAGE DAILY BALANCE OF $341.61 IS	5.12	
09	24	033x	COSMETICS		23.15 CR
09	24	1608	PAYMENT		18.00
09	29	TAMT	WOMENS WEAR 29	38.23	
			SWEATER 43	11.19	

FINANCE CHARGE RATE(S)	ON AVERAGE DAILY BALANCE OF:	MONTHLY PERIODIC RATE	ANNUAL PERCENTAGE RATE
	$.01 TO $33.00	MIN. FINANCE CHARGE $.50	
	$33.01 TO $500.00	1.5%	18%
	$500.00 AND ABOVE	1.25%	15%

use ACCOUNT NO. on correspondence	BILLING DATE	PREVIOUS BALANCE	NEW BALANCE	MINIMUM PAYMENT
K 7605 3821F	OCT 12 19--	345.81	359.20	18.00

To avoid a FINANCE CHARGE next month, pay the NEW BALANCE shown above within 30 days (28 days for February statements) from BILLING DATE. If you prefer to pay in installments pay the MINIMUM PAYMENT shown above, or more, within 30 days (28 days for February statements) from BILLING DATE. The sooner you pay and the more you pay, the smaller your FINANCE CHARGE.

As you can see, Georgette had a previous balance of $345.81. During the month she made purchases of $38.23 and $11.19. She had a finance charge of $5.12. In addition, she made a minimum payment of $18.00, and she returned an item for credit of $23.15. Her new balance is $359.20. Here's how the new balance was computed.

$$\frac{\text{Previous}}{\text{Balance}} + \frac{\text{Finance}}{\text{Charge}} + \frac{\text{New}}{\text{Purchases}} - (\text{Payments} + \text{Credits}) = \frac{\text{New}}{\text{Balance}}$$

$$\$345.81 + \$5.12 + (\$11.19 + \$38.23) - (\$18.00 + \$23.15) = \$359.20$$

Self-Check

Directions: Compute the new balance for the following accounts. Then compare your answers with those in the back of the book.

1. John Littlehawk's monthly credit card statement from Freedom Bank shows a previous balance of $740. During the month he made a payment of $50. There were no new charges. There is a finance charge of $10. What is his new balance? _____

2. Whitney James's monthly statement from Fayette Department Store shows a previous balance of $128.45. She made a payment of $20.00. During the month, she made new purchases of $8.99, $49.99, and $29.99. Her finance charge is $1.90. What is her new balance? _____

3. The monthly statement for Julius Bell from Broadway Men's Store shows a previous balance of $445.37. During the month he made a payment of $200.00 and returned an item for credit of $59.99. He made two purchases of $185.00 and $29.99. His finance charge is $6.59. What is his new balance? _____

Lines of Credit

Most businesses set limits on how much credit they will give customers. That limit, called a **line of credit**, is shown as $1,000 on Georgette's statement on page 20. The line of credit is the maximum amount that Georgette is authorized to charge. Her available credit is $640. **Available credit** is the difference between the line of credit and the account balance. In other words, available credit is the amount Georgette can charge before reaching her credit limit. The "amount due this statement," $18, is the minimum amount that must be paid. Georgette can pay the entire amount if she wishes, but she must pay at least $18.

Self-Check

Directions: Fill in the blanks using the information given. Then compare your answers with those in the back of the book.

	Line of Credit	Account Balance	Available Credit
1.	$1,000	$200	_____
2.	_____	$456	$1,544
3.	$800	$45	_____
4.	$1,200	_____	$56
5.	$500	$238	_____
6.	$3,000	_____	$1,211
7.	$1,500	$522	_____

Practice

Directions: Solve the following problems. Then compare your answers with those in the back of the book.

1. Penny Williams's monthly credit card statement from Lake State Bank shows a previous balance of $39.99. During the month she made a payment of $39.99. There were no new charges or finance charges. What is her new balance? _____

2. David Rubin's monthly statement from Lacy's Department Store shows a previous balance of $287.65. During the month he made a payment of $87.65. He returned two items for credits of $15.99 and $23.99. He made three purchases of $49.00, $8.00, and $699.98. His finance charge is $6.59. What is his new balance? _____

3. Ralph Gomez purchased new furniture for his home. He bought a sofa at $459.99, 2 end tables at $41.99 each, 2 lamps at $25.99 each, and an arm chair at $211.59. There is a 5% sales tax on all items. Figure the following:

   Subtotal: _____

   Sales tax: _____

   Total cost to Ralph: _____

Help with *Shopping for Credit*

Most consumer credit advisers suggest that individuals not spend more than 20%–25% of their income on credit payments, excluding home loans. To avoid paying a high price for credit, keep the following suggestions in mind when shopping for credit:

- Shop for the best source of credit available. For example, a credit union may charge lower interest than a commercial bank.

- Determine which type of credit best suits your needs: a 30-day charge account; a revolving charge account; an installment plan; or bank credit cards.

- Compare interest rates, payment plans, and other service charges from different credit sources to be sure you are getting the best deal. Annual credit card fees commonly range from $20 to $45 a year and interest rates may vary as much as 3%–5%.

- To reduce interest charges, consider paying more than the minimum payment due each month.

- Before you sign any credit contract, be sure to read the fine print! Many contracts may include additional payment items or clauses that you may not be aware of. Don't make that mistake.

Computing Finance Charges

Most businesses do not charge interest for credit cards if the balance is paid within 30 days after the customer receives the statement. If the balance is paid after the 30-day period, however, finance charges are added. Several different methods are used to calculate finance charges, including the previous-balance method, the unpaid-balance method, and the average daily balance method.

Periodic Rate

In all three methods, the finance charge is determined by multiplying a given amount by a **periodic rate**, the monthly finance charge rate. The monthly rate is arrived at by dividing the APR by 12, the number of months in a year. If the APR is 18%, the interest rate is 1.5% per month (18% ÷ 12 = 1.5%). To find the APR, if you know the interest rate per month, multiply the interest rate by 12 (1.5% × 12 = 18%).

Self-Check

Directions: Fill in the blanks using the information given. Then compare your answers with those in the back of the book.

APR	Monthly Interest Rate
1. _____	1.7%
2. 16.44%	_____
3. 12.6%	_____

Previous-Balance Method

Some businesses compute finance charges using the **previous-balance method**. In this method, the finance charges are based on the amount the customer owes as of the closing date of the last billing cycle. This balance is multiplied by the periodic rate to arrive at the finance charge. If you had a previous balance of $1,300.00 and a monthly interest rate of 1.5%, your finance charge would be $19.50 when calculated using the previous-balance method.

Periodic Rate × Previous Balance = Finance Charge
1.5%　　　　　× $1,300.00　　　= $19.50

Self-Check

Directions: Use the previous-balance method of figuring finance charges to solve the following problems. Then compare your answers with those in the back of the book.

1. Brad Fernandez has a charge account with a previous balance of $215.11. The periodic rate is 1.25%. What is the finance charge?

---

2. Carla Loder has a charge account with a periodic rate of 1.6%. Carla's previous balance is $1,255.29. What is the finance charge?

_____

3. Mike Nizel's charge account has a previous balance of $534.56. The periodic rate is 0.75%. What is the finance charge?

_____

Unpaid-Balance Method

Other businesses calculate credit-card finance charges using the **unpaid-balance method**. In this method, rather than use all of the previous balance to compute the finance charge, only the unpaid balance is used. To determine the amount of the unpaid balance, any payments or credits are first subtracted from the previous balance. Here is the formula:

Previous Balance − (Payments + Credits) = Unpaid Balance
Unpaid Balance × Periodic Rate = Finance Charge

EXAMPLE

Your previous balance was $853 and you made a payment of $83. Using the periodic rate 1.2%, the finance charge is $9.24.

Previous Balance − (Payments + Credits) = Unpaid Balance
$853 − $83 = $770
Unpaid Balance × Periodic Rate = Finance Charge
$770 × .012 = $9.24

Self-Check

Directions: Use the unpaid-balance method to solve the following problems. Then compare your answers with those in the back of the book.

1. Sara Goldburn's account statement shows an unpaid balance of $165.90. She made a payment of $25.00. The periodic rate is 1.5%. What is Sara's finance charge? _____2.10_____

2. Julio Tucker's account statement shows an unpaid balance of $1,298.23. He made a payment of $50.00 and received a credit of $32.99. The periodic rate is 1.4%. What is Julio's finance charge?

_____

Average Daily Balance Method

Many banks and retail stores calculate finance charges using the average daily balance method. The **average daily balance** is the average of the account balance at the end of each day of the billing period. New purchases made during the billing period are not included when figuring the balance at the end of the day. Only payments affect the average daily balance at the end of the day. After the average daily balance is computed, the finance charge can be found. Here is the formula:

$$\text{Average Daily Balance} = \frac{\text{Sum of Daily Balances}}{\text{Number of Days}}$$

Brad Highsmith's bank credit card indicates that his last billing period was from June 4 to July 4. His balance on June 4 was $286.15. His bank charges interest of 1.5% per month (18% APR) on the average daily balance. His statement credits him with a payment of $45.00 on June 18. What is his average daily balance for the billing period? What is his finance charge for the billing period?

Step 1 Find the sum of daily balances.

Dates	Payment	End-of-Day Balances		Number of Days	Sum of Balances
6/4–17		$286.15	×	14	$4,006.10
6/18	$45.00	241.15	×	1	241.15
6/19–7/3		241.15	×	15	3,617.25
			Total	30	$7,864.50

Step 2 Find the average daily balance.

Sum of Daily Balances ÷ Total Days = Average Daily Balance
$7,864.50 ÷ 30   = $262.15

Step 3 Compute the finance charge.

Average Daily Balance × Finance Rate (APR) = Finance Charge
$262.15 × 1.5%   = $3.93

The finance charge for Brad's billing period is $3.93.

Self-Check

Directions: Use the average daily balance method to solve the following problems. Then compare your answers with those in the back of the book.

1. Rose Garcia's credit card statement from Designers' Dress Shop is for the billing period of June 1 through June 30. Her previous balance was $300. On June 11 she made a payment of $100. What is the average daily balance of Rose's statement? _____

2. George Randolph's credit card statement for July 1 through July 31 shows a previous balance of $216.44. On July 13 he made a payment of $37.50. What is his average daily balance?

Practice

Directions: Solve the following word problems. Then compare your answers with those in the back of the book.

1. Ken Zucker has a charge account with a periodic rate of 1.2%. Ken's previous balance is $2,349.80. Use the previous-balance method to figure the finance charge. _____

Based on the information given in the table, use the unpaid-balance method to solve the following problems.

	Previous Balance	Payments	Credits	Unpaid Balance	Periodic Rate	Finance Charge
2.	$145.70	$ 16.00	0	_____	1.30%	_____
3.	$ 81.50	$ 25.00	$9.81	_____	1.2%	_____
4.	$792.98	$130.00	0	_____	2.0%	_____

Using a Calculator: *Average Daily Balance*

You can use the memory keys of your calculator to quickly compute the average daily balance. For example, Shari received her charge account statement for the period of April 5 through May 4. On April 5, her balance was $587.50. She made a payment of $50.00 on April 15. What is her average daily balance?

Dates	Payment	End-of-Day Balance		Number of Days		Daily Balances
4/5–4/14		$587.50	×	10	=	$ 5,875.00
4/15	$50.00	537.50	×	1	=	537.50
4/16–5/4		537.50	×	19	=	10,212.50
		Total		30		$16,625.00

1. Find the balance for the first period and put it into memory.
   Press **5 8 7 . 5 0 × 1 0 =**   (5,875 will be displayed)
   Press **M +**   (this puts 5,875 into memory)

2. Find the balance for the second period and put it into memory.
   Press **5 3 7 . 5 0 × 1 =**   (537.50 will be displayed)
   Press **M +**

3. Find the balance for the third period and put it into memory.
   Press **5 3 7 . 5 0 × 1 9 =**   (10,212.50 will be displayed)
   Press **M +**

4. Press **MR**   (16,625 will be displayed)

   All of the subtotals have been stored. By pressing MR, the subtotals are automatically added and totaled. The sum of the daily balance is $16,625.00.

5. Press **÷ 3 0 =**   (554.1666 will be displayed)

   Dividing the number of days (30) into the sum of the daily balances gives you the average daily balance of $554.17.

Unit 8 Review

Developing Your Skill

Directions: Follow the instructions for each set of problems. Based on the the information given, find the new balance.

	Previous Balance	Finance Charge	New Purchases	Payments and Credits	New Balance
1.	$00.00	$0	$62.38	$0	_____
2.	$25.00	$2.34	$167.97	$18	_____
3.	$600	$10.50	$107.74	$65	_____
4.	$1,000	$15	$0	$25	_____
5.	$1,500	$24	$257.34	$45	_____

Using the information given, find the sum of daily balances, the average daily balance, and the finance charge for the following.

	End of Day Balance	Number of Days	Sum of Daily Balances	Average Daily Balance	Monthly Interest Rate	Finance Charge
6.	$84.70	15	_____		1.5%	
	$66.70	1	_____			
	$66.70	14	_____	_____		_____
7.	$238.16	16	_____		1.25%	
	$193.16	1	_____			
	$193.16	15	_____	_____		_____

Solve the following problems for installment loans. Fill in the missing blanks in the columns. When necessary, round numbers to the nearest cent. Use the tables on pages 116 and 117.

	Price	Percent Down Payment	Amount Financed	Term (Months)	APR	Monthly Payment	Total Repaid	Finance Charge
8.	$2,000	25%	_____	18	15%	_____	_____	_____
9.	$1,500	0%	_____	24	15%	_____	_____	_____
10.	$600	$16\frac{2}{3}$%	_____	6	18%	_____	_____	_____

Directions: In solving a mathematics word problem, ask: (1) What information is being asked for? (2) What information is given in the problem? (3) What information is needed to solve the problem? Use the formulas in this unit to solve these problems.

11. David Goldberg's bank card statement shows a previous balance of $800. His finance charge is $12. He made no new purchases. He made a payment of $45. What is David's new balance?

   _____

12. Chris Brown's credit card statement shows a previous balance of $49.95. Her finance charge is $1.00. She made new purchases of $19.95, $29.95, $75.98, and $250.00. She made a payment of $15.00. What is her new balance?

   _____

13. Lola Martinez' bank card statement shows a previous balance of $457.19. She made a payment of $50.00 She made new purchases totaling $37.17, $29.00, and $65.99. Her finance charge is $8.23. What is the new balance of Lola's account?

   _____

14. Pat Chang's credit card statement shows a previous balance of $123.66. He made a payment of $15.00. He had one new purchase for $34.99. He returned one item in the amount of $33.99. His finance charge was $1.85. What was Pat's new balance?

   _____

15. Paul Oliver has a charge account that uses the average daily balance method for calculating finance charges. His billing period is June 7 through July 6. His previous balance was $198.33. He made a payment of $18.00 on June 10. What was his average daily balance for the month?

   _____

16. Kathy Derwitz opened a charge account at Clancy's Department Store with the purchase of a stereo costing $675. The monthly rate of interest for the finance charge is 1.5%. The billing period is December 1 through December 31. On December 15 Kathy made a payment of $50.00. What was Kathy's average daily balance for the month? _____

   What was the finance charge for the billing period?

   _____

Figuring Discounts and Markups

Some stores always seem to offer the best deals in town. How can these stores charge less than their competitors for the goods they sell and still make a profit? Certain stores take advantage of special trade discounts and actually pay less for the goods and then pass the savings on to their customers. Other stores use markups to set prices that can both attract customers and return a fair profit.

Figuring Trade Discounts

Most products you buy have traveled through a channel of distribution that includes manufacturers, wholesalers, and retailers. Manufacturers usually do not sell their products directly to you, the final customer. Rather, the manufacturers usually deal with wholesalers, who in turn resell the products to retail businesses and stores.

Retailers often buy items through wholesalers' catalogs. In the catalogs, the price shown for an item is the **catalog price**, or the **list price**. This is the price the general public would pay. The retailer, however, buys the item at a **trade discount**, which is a discount from the catalog price.

Finding the Net Price

The trade discount is often expressed as a percent of the list price. This percent is called the **trade-discount rate**. The price the business actually pays for the item after the discount is taken is referred to as the **net price**.

To find the net price of an item, first you must determine the amount of the trade discount by multiplying the trade-discount rate by the list price. Then subtract the trade discount from the list price as follows:

Trade-Discount Rate × List Price   = Trade Discount
List Price − Trade Discount = Net Price

EXAMPLE   Dan Matheny Electronics Company allows a 40% trade discount on any 19-inch television set listed in its catalog. How much is the trade discount if the set has a list price of $500? What is the net price?

Trade-Discount Rate × List Price   = Trade Discount
40% × $500   = $200
List Price − Trade Discount = Net Price
$500 − $200   = $300

Self-Check

Directions: Use the formula for determining the net price to solve these problems. Then compare your answers with those in the back of the book.

1. Krista Tamaja works at Beacon Automotive and receives a 25% trade discount on a car purchase. Krista wants to purchase a car that has a list price of $10,041. What is Krista's net price after her trade discount?_____

2. Splash Sporting Goods buys skateboards at a list price of $50 each with a 15% trade discount. What is the net price?

_____

3. Jodi Bernstein works part-time at the school bookstore. Jodi receives a 20% trade discount on all book purchases. If Jodi purchases a book that costs $40, what is the net price Jodi will pay?

_____

The Complement Method

The **complement method** is a short cut to determine the net price. To use the complement method, you first subtract the discount rate from the full rate (100%). The result is the **complement**, or the percent that you actually pay. Then, multiply the complement by the list price to find the net price.

EXAMPLE

The Singer Company sells telephones at a list price of $139. If there is a trade discount of 25%, what is the net price of the telephones?

Step 1 Full Rate − Discount Rate = Complement
100% − 25% = 75%

Step 2 Complement × List Price = Net Price
75% × $139.00 = $104.25

Self-Check

Directions: Solve the following problems using the complement method. Then compare your answers with those in the back of the book.

1. Gregorio's Appliances bought a TV that had a list price of $299.50 with a 10% trade-discount rate. What was the net price?

   _____

2. Maurice Rizzo belongs to the Bionic Bicycle Club and receives a 25% discount at Cycle Sports for all purchases. Maurice is planning to purchase a new bicycle that lists for $400. How much will the net price be after the trade discount? _____

Finding the Trade-Discount Rate

Some businesses put only the list prices and net prices in their catalogs. They do not give the trade-discount rate. However, when both prices are known, the trade-discount rate can be determined. Trade discount rate can be determined by subtracting net price from list price, then dividing trade discount by list price.

EXAMPLE

Granite Stoves lists a wood stove in its catalog for $420, with a net price of $280. What is the trade-discount rate?

List Price − Net Price = Trade Discount
$420 − $280 = $140
Trade Discount ÷ List Price = Trade-Discount Rate
$140 ÷ $420 = $33\frac{1}{3}$%

Self-Check

Directions: Use the formula for determining the trade-discount rate to solve the following problems. Then compare your answers with those in the back of the book.

1. The Belknap Wholesale catalog has a list price of $28.50 for a set of wrenches. The net price after the trade discount is $21.37. What is the trade-discount rate? _____

2. The Sparkling Pan Kitchen Shop catalog features food processors with a list price of $150.00. The net price after the trade discount is $135.00. What is the trade-discount rate? _____

3. The Photography Warehouse sells Beta Zoom lenses at a list price of $172.50. The net price after the trade discount is $125.50. What is the trade-discount rate? _____

Practice

Directions: Solve the following problems. Then compare your answers with those in the back of the book.

1. Randy works at a fashion boutique and receives a 34% discount on all purchases. Randy wants to purchase a leather coat with a list price of $200. After his trade discount, what is the net price Randy will pay for his coat? _____

2. At an auto parts store, the list price for a set of hubcaps is $400 and the net price is $250. What is the trade discount?

_____

3. In the Brill Factory Warehouse advertisement for shoes, the prices before and after the trade discount are listed. Determine the amount of the trade discount and the trade-discount rate for each.

	List Price	Net Price	Trade Discount	Trade-Discount Rate
Men's Jogging Shoes	$47.98	$38.38	_____	_____
Ladies' Boat Shoes	$35.55	$31.28	_____	_____
Boys' Sneakers	$29.99	$25.50	_____	_____

Figuring Cash Discounts

All businesses would rather have their customers pay sooner than later. To encourage early payment, many businesses offer their customers a choice of two prices: (1) the regular total price of the products, or (2) a lower price *if* the bill is paid early. Here's how it works.

A business sends a customer an **invoice**, which is another name for the bill. The total price of the products that the customer bought is shown on the invoice. Also shown are the terms of getting the lower price. The terms of the lower price are shown as a percentage of decrease off the invoice price. The customer can get the lower price, however, only if the bill is paid within a certain number of days.

The reduction off the invoice price is called a **cash discount**. When the cash discount is expressed as a percent, it is called the **cash-discount rate**. If the customer meets the terms and the cash discount is taken, the final, lower price is called the **cash price**. If no discount is taken within the discount period, the net price, or full amount, has to be paid. On the invoice, the terms of a cash discount are generally shown as follows:

2/10, n/30

This notation means that a 2% discount will be allowed if the invoice is paid within 10 days of the date on the invoice. If the invoice is paid any time *after* 10 days, the net price (n) must be paid. In any event, the invoice must be paid within 30 days. Look at the invoice that follows. The cash-discount rate is 2%.

PLAZA SPORTS			Invoice	

655 EAST MEADOW ROAD   RALEIGH, NC 27611 PHONE: 919-555-1354

TO:

```
┌                            ┐
  Thorson Ski Shop
  2600 Clearview Blvd.
  Minneapolis, MN 55412
└                            ┘
```

Date: 3/14/--
Invoice No.: 2-68-67
Order No.: AD 2301
Shipped By: Air Freight
Terms: 2/10, n/30

QUANTITY	ITEM	UNIT PRICE	TOTAL
3	Bengton Skis (sets)	$123.50	$370.50

1. What is the last day of the discount period? *March 24.*

2. When can a cash discount be taken? *Only if payment is made within the discount period.*

3. When must the net amount of the invoice be paid, if not paid within the discount period? *March 14 + 30 days, or April 13, since March has 31 days.*

To determine the cash-discount amount, multiply the cash-discount rate by the net price. To find the cash price, subtract the cash discount from the net price.

$$\text{Cash-Discount Rate} \times \text{Net Price} = \text{Cash Discount}$$
$$\text{Net Price} - \text{Cash Discount} = \text{Cash Price}$$

Look again at the invoice. This is how the cash price was determined.

$$\text{Discount Rate} \times \text{Net Price} = \text{Cash Discount}$$
$$2\% \times \$370.50 = \$7.41$$
$$\text{Net Price} - \text{Cash Discount} = \text{Cash Price}$$
$$\$370.50 - \$7.41 = \$363.09$$

Occasionally, more than one cash discount is offered. For example, when an invoice offers 2/10, 1/15, n/30, a 2% cash discount can be taken if the invoice is paid within 10 days. A 1% cash discount can be taken if it's paid within 15 days. The net amount of the bill must be paid within 30 days.

EXAMPLE

The Royal Gate Fence Company bought $2,285 worth of fencing from Home Manufacturers, Inc. The invoice is dated May 17. The terms are 3/10, 2/20, n/30. What is the cash price for each discount?

a. What is the last date the 3% discount may be taken? *May 27*
What is the 3% discount? *3% × $2,285.00 = $68.55*
What is the cash price after the 3% discount?
$2,285.00 − $68.55 = $2,216.45

b. What is the last date the 2% discount may be taken? *June 6*
What is the 2% discount? *2% × $2,285.00 = $45.70*
What is the cash price after the 2% discount?
$2,285.00 − $45.70 = $2,239.30

There are several differences between trade and cash discounts. Cash discounts are taken off the invoice price. Trade discounts are taken off the catalog price. To get a cash discount, the customer must pay early. To get a trade discount, the customer doesn't have to meet any conditions. Cash discounts are the same for all buyers. Different trade discounts may be offered to different customers. For instance, a retailer who orders many items may get 40% off the catalog price. A smaller retailer who orders just a few items may get only 5% off the catalog price.

Self-Check

Directions: Use the formulas for determining the cash discount and/or the cash price to solve these problems. Then compare your answers with those in the back of the book.

1. Winkelman's Nursery received an invoice dated February 6 for $950.00. Terms are 3/10, n/30. The manager of Winkelman's pays all bills on the 15th of the month. For what amount will he write the check to pay the invoice? _____

2. On October 20, Hardware Distributors sold goods to Pahl Hardware that totaled $1,574.80. Terms of the invoice were 2/10, n/30. If the invoice was paid on October 29, what was the cash price? _____

3. Prescott Department Store received an invoice dated March 24 for an order of dishes totaling $3,875.00. Terms were 3/15, n/30. The invoice was paid on April 6. What was the cash price?

_____

4. Pitcairn Gift Shop received an invoice dated August 28 for $768.00 with terms of 2/10, n/30. The invoice was paid on September 5. What is the cash price? _____

5. The Craft Shop received an invoice dated June 15 for $428.34. The terms were 2/10, 1/20, n/30. What is the latest date each cash discount can be taken, and what will be the cash price on each date? _____

Practice

Directions: Use the formulas for determining the cash discount and the cash price to solve these problems. Then compare your answers with those in the back of the book.

1. On October 28, Athletic Distributors sent an invoice to the Pro Shop for $2,674.95. Terms are 3/10, n/30. When the Pro Shop paid the invoice on November 4, what amount did it pay?

_____

2. On April 29, Softcover Books received an invoice for $864.22 with terms of 2/10, n/30. If all bills are paid on the 5th of the month, for what amount will the bookkeeper make out the check?

_____

3. Express Travel Bureau received a $562.85 invoice for printed materials on November 22 with terms of 3/15, n/30. What was the amount of the check that was mailed on December 3?

_____

4. On October 19, Circle Hardware sent an invoice to Country Apartments for $3,120.89. Terms were 2/10, n/30. Country Apartments paid the invoice on October 25. What amount was paid on that date? _____

5. Ken's Auto Store received an invoice from a distributor on August 28 for $2,690.00. Terms of the invoice were 3/15, n/30. What amount did Ken's Auto pay on September 28? _____

Determining Chain Discounts

In some cases, more than one trade discount is given. For example, in addition to the normal catalog discount, a manufacturer may want to offer additional discounts to further lower the price of discontinued products or to encourage large orders. This practice is referred to as a multiple or **chain discount**. These discounts, which are really a series of discounts, may be expressed on an invoice as as 20/10/5, which means "20% less 10% less 5%."

EXAMPLE   A manufacturer of trailers is closing out a model that has not sold well in spite of previous discounts. The trailer is priced at $4,000. A chain discount of 20/10/5 is now offered. Find the net price.

Step 1   Find the first discount.   20% × $4,000 = $800
Find the first net price. $4,000 − $800   = $3,200

Step 2   Find the second discount.   10% × $3,200 = $320
Find the second net price. $3,200 − $320   = $2,880

Step 3   Find the third discount.   5% × $2,880 = $144
Find the third net price. $2,880 − $144   = $2,736
The final net price is $2,736.

Notice that at each step, the discount is taken from the previous net price. You *cannot*, however, add up the discounts and get the correct answer.

EXAMPLE   Using the same trailer price of $4,000 and chain discount of 20/10/5, see what happens when you try to add up the discounts.

20% + 10% + 5% = 35%
35% × $4,000   = $1,400
$4,000 − $1,400   = $2,600

As you can see, when you compare this answer with the correct one of $2,736, this method leads to the wrong answer.
You can, though, use the complement method, following these steps.

EXAMPLE   Again, using the same price and chain discount, find the net price by means of the complement method.

Step 1   Find the complements of the chain discounts.
The complement of 20% = 80%
The complement of 10% = 90%
The complement of 5% = 95%

Step 2   Find the discount rate.
80% × 90% × 95% = 68.4%

Step 3   Find the net price.
68.4% × $4,000 = $2,736

Self-Check

Directions: Use the steps for chain discounts to solve these problems. Remember that the complement method is a short cut! Then compare your answers with those in the back of the book.

1. A desktop copier from Sutherland Wholesalers has a catalog price of $439.75 with a 15/10/5 chain discount. What is the net price?

    _____

2. Lipscomb Company has an oversupply of some of its less popular light fixtures. What is the net price of each item?

Stock Number	List Price	Chain Discount	Net Price
311AO	$79.95	30/15/10	_____
4086LN	$55.55	20/10/5	_____
118XT	$123.85	35/25/15	_____

Practice

Directions: Use the formulas for chain discounts to solve these problems. Remember that the complement method is a short cut! Then compare your answers with those in the back of the book.

1. Beauty Supply, Inc., is trying to sell its older model hair dryers at a chain discount of 35/25/15 off the catalog price of $365.95. What is the net price? _____

2. User-Friendly Computers is offering a 20/10/5 chain discount on its $600.00 home computer. What is the net price?

    _____

3. Precision Office Supplies is offering retailers a chain discount of 15/10/5 on its $49.95 adding machines. What is the net price?

    _____

Using a Calculator: *Chain Discounts*

Using a calculator when figuring chain discounts makes it very easy to determine the net price for an item. Let's say you have an item with a list price of $675.00, less discounts of 15/10/5 (15%, 10%, 5%).

To determine the net price, press 6 7 5 − 1 5 % − 1 0 % − 5 %. The calculator automatically determines the amount of the discount for each percent and subtracts that amount from the price.

In this example, the net price is $490.55625. Rounded to the nearest cent, the net price is $490.56.

Figuring Markups

The amount of money for which a business sells its products is called the **selling price**, or the **retail price** if sold by a retailer. The amount of money that a business pays for a product is called its **cost**. All businesses need to establish a selling price for a product that will (1) cover the cost of the product; (2) pay for such expenses as wages, rent, taxes, and insurance; and (3) still allow them to make a profit. The difference between cost and selling price is called the **markup**, determined like this.

Selling Price − Cost = Markup

EXAMPLE   Schultz Clothing purchases a suit for $139.50 and sells it for $219.00. Find the markup amount.

Selling Price − Cost = Markup
$219.00 − $139.50 = $79.50

Self-Check

Directions: Use the formula to find the markup amount to solve these problems. Compare your answers with those in the back of the book.

1. The Fresh Springs Water Distributor purchases bottled water for $.40 cents a gallon and sells it for $1.19 a gallon. Find the markup amount. _____

2. The Fruit Basket grocery store purchases red apples at $.10 apiece and sells them for $.53 apiece. Find the markup amount.

   _____

3. A car dealer purchases a car from the manufacturer for $9,800 and sells it for $12,000. What is the markup amount?

   _____

Markup Rate Based on Selling Price

Markup also can be expressed as a percent of the cost or the selling price. When stated as a percent, the markup is called a **markup rate**. Most businesses use the selling price as the base for their markup rate. When the markup amount and the selling price are known, use this formula.

Markup ÷ Selling Price = Markup Rate Based on Selling Price

EXAMPLE   Karla's Hardware Store buys plastic switchplates for $.25 apiece and sells them for $.69 apiece. What is the markup rate?

Selling Price − Cost = Markup
$.69 − $.25 = $.44
Markup ÷ Selling Price = Markup Rate
$.44 ÷ $.69 = .638 or 64%

Self-Check

Directions: In the problems that follow, the selling price and the cost for various items are given. Find the markup and the markup rate based on selling price. Then compare your answers with those in the back of the book.

Selling Price	Cost	Markup	Markup Rate
1. $79.50	$26.50	_____	_____
2. $5,700.00	$2,540.00	_____	_____
3. $98.88	$74.16	_____	_____

Markup Rate Based on Cost

Some businesses, such as grocery stores, determine the markup rate on the basis of cost. When the markup amount and cost are known, use this formula.

Markup ÷ Cost = Markup Rate Based on Cost

EXAMPLE

Foodway Stores buys grapes at $.25 per pound. It sells them for $.75 per pound. What is the markup rate based on cost?

Selling Price − Cost = Markup
$.75 − $.25 = $.50
Markup ÷ Cost = Markup Rate
$.50 ÷ $.25 = 2, or 200%

EXAMPLE

Jorgenson Creamery Stores buys milk from a wholesaler for $1.99 per gallon and sells it for $2.09 per gallon. What is the markup rate based on cost?

Selling Price − Cost = Markup
$2.09 − $1.99 = $.10
Markup ÷ Cost = Markup Rate
$.10 ÷ $1.99 = 5%

Self-Check

Directions: In the problems that follow, the selling price and the cost for various items are given. Find the markup and the markup rate based on cost. Then compare your answers with those in the back of the book.

Selling Price	Cost	Markup	Markup Rate
1. $1.50	$1.30	_____	_____
2. $.69	$.35	_____	_____
3. $5.99	$5.16	_____	_____
4. $2.98	$2.59	_____	_____
5. $.39	$.29	_____	_____

Practice

Directions: In the following problems, the selling price and the cost for various items are given. Find the markup and the markup rate based on selling price. Then compare your answers with those in the back of the book.

	Selling Price	Cost	Markup	Markup Rate
1.	$60.45	$39.00	_____	_____
2.	$14.59	$10.95	_____	_____
3.	$108.00	$27.50	_____	_____

In the following problems, determine the markup and the markup rate based on cost.

	Selling Price	Cost	Markup	Markup Rate
4.	$.42	$ .22	_____	_____
5.	$12.25	$10.95	_____	_____
6.	$.85	$ .62	_____	_____

Help with *Estimating Discounts and Markups*

Estimating answers can be very helpful when working with discounts and markups. If the actual answer is very different from your estimated answer, you'll know you made a mistake in your calculations.

To estimate an answer, you might want to change a percent to a fraction. For example, to determine a $33\frac{1}{3}$% discount on a sweater costing $35.95, change $33\frac{1}{3}$% to the fraction $\frac{1}{3}$. Round the cost of the sweater to $36. You know that $\frac{1}{3}$ of $36 is $12. So, $12 is the approximate amount of the discount. To estimate the net price, subtract the discount from the cost: $36 − $12 = $24. You also can estimate the net price by figuring that if $\frac{1}{3}$ (the $12 discount) has been taken away, you have $\frac{2}{3}$ or $24 (2 × $12) remaining.

Another method to use when estimating discounts is to first find 10% of a number by moving the decimal point one place to the left: 10% of $89.90 = $8.99. To find 30% of $89.90, move the decimal point one place to the left, round the amount to $9, and multiply by 3 (3 × 10% = 30%). Therefore, 30% of $89.90 is approximately $27 ($9 × 3 = $27).

Rounding can be used for estimating markups, too. For example, suppose you want to mark up an $8.95 item by 400%. Round $8.95 to $9.00. Change 400% to a whole number by dropping the percent sign and moving the decimal point two places to the left. The whole number is 4. Then, multiply $9 × 4 to get $36, the estimated markup.

Unit 9 Review

Developing Your Skills

Directions: In the tables that follow, all the information that you will need to solve each problem is given. Use the formulas that you have learned in this unit. Round all figures to the nearest cent.

	Date of Invoice	Terms	Paid	Net Price	Cash Discount	Cash Price
1.	7/23	2/10, n/30	8/1	$630.50	_____	_____
2.	1/24	3/10, n/30	2/4	$214.75	_____	_____
3.	6/15	4/7, n/30	6/24	$750.00	_____	_____
4.	11/30	3/10, n/30	12/9	$980.20	_____	_____
5.	3/1	1/15, n/30	3/10	$2,456.00	_____	_____

	List Price	Trade-Discount Rate	Complement	Net Price
6.	$156.90	13%	_____	_____
7.	$240.00	$33\frac{1}{3}\%$	_____	_____
8.	$.70	25%	_____	_____
9.	$380.98	16%	_____	_____
10.	$110.50	21%	_____	_____

	Cost	Markup Rate Based on Cost	Selling Price
11.	$11.20	$33\frac{1}{3}\%$	_____
12.	$1,625.00	20%	_____
13.	$44.29	75%	_____
14.	$150.45	46%	_____
15.	$23.80	50%	_____
16.	$.59	5%	_____
17.	$.39	10%	_____

Directions: In solving a mathematics word problem, ask: (1) What information is being asked for? (2) What information is given in the problem? (3) What information is needed to solve the problem? Use the formulas in this unit to solve these problems.

18. Ryerson Stores buys hair dryers at $18 each and has a 20% markup based on the past selling price. What is the selling price of the hair dryers? _____

19. On an invoice dated 7/30, Appleby & Company was billed for merchandise costing $436, terms 2/5, n/30. The bill was paid on 8/2. What is the cash price? _____

20. Brosky Mail Order House has a family-sized camping tent kit in its catalog with a list price of $356.00 and a net price of $295.48. What is the trade-discount rate? _____

21. A sun lamp has a price of $25.60. Wright Department Store marks the sun lamp up to $32.00. What is the markup rate based on cost?

_____

22. The Fulbright Manufacturing Company is overstocked on some playground sets on which it has a suggested list price of $184.75 and a chain discount of 15/10/5. Find the net price.

_____

23. Home Furnishings, Inc., buys garage door openers for $200.96 and sells them for $314.00. What is the markup rate based on cost?

_____

24. Baker Pharmacy buys one brand of cologne for $2.64 and sells it for $6.00. What is the markup rate based on the selling price?

_____

25. West Coast Fabrics offers the following chain discounts to wholesalers: 20% on orders over $5,000, 20% less 10% on orders over $10,000, and 20% less 10% less 5% on orders over $15,000. A wholesaler places an order for $20,000 worth of fabric to get the full 20/10/5 discount. What is the net price of the order?

_____

Posttest

Directions: Follow the instructions for each set of problems.

Solve the following problems.

1. $65.79 + 35.8 =$ _____

2. $\$7,839 - \$492 =$ _____

3. $11,568 \times 1.25 =$ _____

4. $765\overline{)42,740}$

Change the following written amounts to figures.

5. Two hundred nine _____

6. Sixty million, thirty-four thousand, eight hundred one _____

Round each number to the underlined place.

7. 8,<u>9</u>75,604 _____

8. 113,<u>5</u>91 _____

9. <u>5</u>25 _____

10. 21,<u>5</u>01 _____

Solve the following problems. Reduce answers to lowest terms.

11. $\frac{8}{11} + \frac{1}{3} + \frac{2}{9} =$ _____

12. $15\frac{3}{8} - 7\frac{2}{3} =$ _____

13. $3\frac{1}{8} \times 2\frac{5}{6} =$ _____

14. $7.49\overline{)61.418}$

15. $77.775 + 9.938 + 7.9 =$ _____

16. $122.31 \times .06 =$ _____

Reduce these fractions to their lowest terms.

17. $\frac{36}{12}$ _____

18. $\frac{80}{8}$ _____

19. $\frac{135}{9}$ _____

20. $\frac{64}{4}$ _____

Solve the following word problems.

21. Beth scored 90% on a business math test. If she answered 135 questions correctly, how many questions were on the test?

_____

22. The interest on Roy Price's 9-month loan was $367.50. If the interest rate was 7%, what was the amount of the principal?

_____

23. Three people agreed to share equally the profits of a video store. The store earned the following amounts: June, $2,480.12; July, $3,210.50; and August, $4,814.76. How much did each person earn during those months?

_____

24. David Pulaski is paid $8.50 per hour for 40 hours and time-and-a-half for all hours over 40. He worked 48 hours and had the following deductions: federal tax, 22%; state tax, 3%; city tax, 5%; FICA, 7.51%. What is his net pay?

_____

25. Harriet's checking account balance was $486.22. She wrote the following checks: $48.26, $20.20, $180.40, $210.89, $7.39, and $99.81. She also made deposits of $50.00, $210.00, and $386.20 and had a miscellaneous check charge for $7.10. What was her new balance?

_____

26. Mid-Central Supply offers chain discounts of 20%, 10%, and 5% on all purchases. If someone purchased furniture costing $12,000 what would the net price be?

_____

27. Randolph Tires pays $65.00 per tire for top-quality tires. The company needs a markup of 35% over cost to make a profit. What should be the selling price of the tires?

_____

28. Jutoy bought a car for $7,800. Her down payment was $800 and she agreed to pay the loan back in 36 monthly payments of $215.83. What is the amount of interest she will pay?

_____

29. Jan Oliver worked a total of 208 hours over 4 weeks. Her regular pay is $7.50 per hour. She is paid time-and-a-half for all hours over 40 each week. What was her total pay for the 4 weeks?

_____

30. Tri-County Sales gives customers a cash discount of 2/15, n/30. If a customer receives a bill dated January 17 for $900, what is the amount due if the bill is paid on February 7?

_____

31. Malcolm is paid on a piecework basis of $.37 per piece. On Monday he completed 200 pieces; Tuesday, 177; Wednesday, 192; Thursday, 180; Friday, 178. What was his gross pay for the week?

_____

32. Benson's Ice Cream Company raised the price of its ice cream cones from $1.10 to $1.25. What was the percent of increase?

_____

33. Sherry is paid a weekly salary of $425 plus a commission of 12% on all sales over $10,000. If Sherry had $17,000 sales for the week, what was her gross pay for the week?

_____

34. During a recent football game, Kendall made 15 passes. Of these, only 6 were completed. What was the percent of passes completed?

_____

35. Bette took a loan of $2,500 at a rate of 11% simple interest. If she repays the loan in 12 months, how much interest will she pay?

_____

Answer Key

Unit 1

Self-Check, pp. 2–3

1.	77	2.	60
3.	1,042	4.	921
5.	812	6.	1,220
7.	930	8.	6,109
9.	10,878	10.	10,899
11.	8,987	12.	3,992
13.	3,798	14.	9,912
15.	12,318	16.	16,873

Self-Check, pp. 3–4

1.	4.595	2.	966.77
3.	1,283.905	4.	9.565
5.	1.9811	6.	352.2089
7.	10.119	8.	40.99
9.	9.9997	10.	478.99
11.	999.99	12.	801.12
13.	1,064.683	14.	274.458
15.	1.2413	16.	177.69

Practice, p. 4

1.	6,777	2.	10,560
3.	14,953	4.	57,559
5.	33,587	6.	58,799
7.	87,481	8.	12,049
9.	62.401	10.	1.909

Self-Check, pp. 5–6

1.	12	2.	19
3.	24	4.	442
5.	213	6.	3,311
7.	649	8.	298
9.	5,893	10.	1,998

Self-Check, p. 6

1.	7.1	2.	4,442.1
3.	.6617	4.	26,214.7
5.	3.054	6.	80.6
7.	2,095.416	8.	71.891
9.	4.89	10.	50.95

Practice, p. 7

1.	2,983	2.	1,759
3.	1,568	4.	106
5.	9,042	6.	8,190
7.	2,901	8.	2,419
9.	16,526	10.	16.51
11.	5.09	12.	20.89

Self-Check, p. 8

1.	1,196	2.	4,712
3.	3,053	4.	3,071
5.	265,515	6.	181,746
7.	60,630	8.	28,182
9.	385,664	10.	259,680
11.	249,527	12.	91,056
13.	31,128,003	14.	3,352,947
15.	1,662,024	16.	1,674,265

Self-Check, p. 9

1.	730	2.	1,220
3.	3,450	4.	9,180
5.	6,100	6.	84,600
7.	100,600	8.	207,600
9.	631,000	10.	186,000
11.	7,210,000	12.	3,080,000

Self-Check, p. 10

1.	.0573	2.	.4968
3.	.00054	4.	.4464
5.	7.92894	6.	317.7999
7.	.9999	8.	.038396
9.	60.63645	10.	.90712

Practice, p. 10

1.	1,533	2.	924
3.	87,344	4.	728,586
5.	4,330,143	6.	1,662,024
7.	73,450	8.	288,900
9.	5,486,000	10.	12.54

Self-Check, p. 12

1.	$53\frac{1}{2}$	2.	$88\frac{1}{8}$
3.	63	4.	$14\frac{1}{15}$

5.	1,465	6.	$34\frac{3}{22}$
7.	23	8.	110
9.	$196\frac{19}{42}$	10.	$76\frac{13}{92}$
11.	381	12.	$50\frac{13}{51}$
13.	253	14.	54
15.	38	16.	$52\frac{225}{622}$

Self-Check, pp. 13–14

1.	4.6	2.	5.71
3.	17	4.	6.5
5.	1.2	6.	15.3
7.	$7.25	8.	$1.20
9.	$1.58	10.	.49
11.	8.788	12.	5.825
13.	6.3846	14.	9.96458
15.	.743256	16.	.086432

Practice, p. 14

1.	$20\frac{3}{8}$	2.	23
3.	$316.46	4.	540
5.	$84\frac{8}{93}$	6.	34
7.	3.5	8.	245
9.	620	10.	2.35
11.	450	12.	$36.50
13.	53.72	14.	36.71
15.	57.6221	16.	211.695

Unit 2

Self-Check, p. 21

	Place	Value
1.	hundreds	400
2.	tens	20
3.	hundredths	$\frac{8}{100}$
4.	thousands	3,000
5.	tenths	$\frac{4}{10}$
6.	thousandths	$\frac{7}{1,000}$

Self-Check, p. 21

1. $10,991.50
2. 10,991.5

3. $465,048.46
4. 1,600,048.09

Self-Check, p. 22

1. sixty-six and nineteen hundredths
2. four hundred nine dollars and thirty-two cents
3. twenty-five thousand, six hundred fifty-nine
4. one thousand, two hundred forty-two dollars and nine cents
5. five hundred ninety-two thousand, one hundred twenty-two
6. twenty-four million, nine hundred fourteen thousand, two hundred seventy-nine dollars and seventy-five cents

Self-Check, p. 22

1.	2.	x	3.	x
4.	5.	x	6.	x

Practice, pp. 22–23

	Place	Value
1.	thousandths	$\frac{4}{1000}$
2.	thousands	3,000
3.	hundreds	600
4.	millions	1,000,000
5.	thousands	7,000
6.	tens	80
7.	tenths	$\frac{4}{10}$
8.	hundredths	$\frac{9}{100}$
9.	hundred thousands	700,000
10.	ones	4

11. four thousand, nine hundred three dollars and fifty-four cents
12. twenty-eight thousand, nine hundred ninety-six dollars and thirty-two cents
13. one hundred two thousand, four hundred sixty-two
14. four million, two hundred one thousand, one hundred ninety-seven
15. six hundred forty-one million, two hundred three thousand, eight hundred thirty-seven and ninety-three thousands

16. 0.005   17. 100.1
18. 6.9   19. $5,065.13
20. 10,000.2

Self-Check, pp. 24–25

1.	310	2.	20
3.	1,680	4.	8,610

5.	50	6.	440
7.	900	8.	123,500
9.	2,900	10.	13,500
11.	1,000	12.	261,400
13.	2,000	14.	2,502,000
15.	100,000	16.	236,000
17.	200,000	18.	10,000
19.	13,000	20.	1,000
21.	261,000	22.	4,000,000
23.	23,000,000		
24.	100,000,000		
25.	10,000,000		
26.	7,000,000		
27.	1,111,000,000		

Self-Check, p. 25

1.	1.8	2.	128.3
3.	62.9	4.	.1
5.	.7	6.	.2
7.	.25	8.	4.12
9.	5.31	10.	26.08
11.	539.35	12.	23.00
13.	93.654	14.	3,432.456
15.	10.333	16.	1.343
17.	4.234	18.	4,324.000

Self-Check, p. 26

1.	$34.33	2.	$5,342.54
3.	$5.56	4.	$99.00
5.	$3.91	6.	$7.35
7.	$78.33	8.	$112.21
9.	$75.36	10.	$322.00
11.	$7.00	12.	$44,567.00
13.	$110.00	14.	$90.00
15.	$678.00	16.	$24.00
17.	$566.00	18.	$7,009.00

Practice, pp. 26–27

1.	$14.89	2.	$5.79
3.	$90.46	4.	$1.23
5.	$35.35	6.	$56.32
7.	$46.00	8.	$36.00
9.	$123.00	10.	$567.00
11.	$90.00	12.	$427.00

Self-Check, p. 27

1.	3,465	2.	15,211
3.	543,245	4.	232,521
5.	654	6.	11,115
7.	3,579	8.	18,431
9.	532,411		

Self-Check, p. 28

1.	319.62	2.	29.794
3.	710.594	4.	0.05
5.	19.131	6.	0.764

Practice, p. 28

1.	43	2.	876	3.	3,510
4.	5,927	5.	3.87	6.	41.6

7. 1.36, 1.47, 1.5
8. 5.09, 5.10, 5.13
9. 6.08, 6.38, 6.81
10. 4.81, 4.86, 4.95

Self-Check, p. 29

	Estimate	Answer
1.	110	108
2.	100	99
3.	870	868
4.	240	235
5.	1,100	1,115
6.	800	798

Self-Check, p. 30

1.	300	332
2.	3,000	2,970
3.	300	270
4.	31,000	30,896
5.	19,000	18,905
6.	20,000	20,915

Self-Check, p. 31

1.	700	792
2.	600	792
3.	2,400	2,408
4.	1,600	1,725
5.	4,500	4,002
6.	3,900	3,668

Self-Check, p. 32

	Estimate	Actual Answers
1.	15	14.21
2.	25	23.77
3.	5	5.21
4.	7.5	7.71
5.	8	8.67
6.	245	305

Practice, p. 32

	Estimate	Answer
1.	5,000	4,912
2.	80,000	80,687
3.	3,700	3,690
4.	2,000	1,967
5.	29,400	25,506
6.	7,000	8,532
7.	138,000	155,278
8.	6	6.08
9.	80	85.72
10.	30	30.29

Self-Check, p. 34

1. $960   2. $2,700   3. $133

Practice, p. 35

1. Baltimore: approximately 10,000 more

2. $140
3. 554
4. $760
5. Peru: 18,700,000
   Bolivia: 6,100,000
   Colombia: 28,600,000
   Brazil: 129,700,000
   Paraguay: 3,100,000

Self-Check, p. 36

1. 24 weeks
2. $200.00
3. $22,800.00

Self-Check, p. 37

1. 3 bags
2. $\frac{3}{4}$ of a pizza apiece
3. 14 feet

Self-Check, p. 38

1. 33 hours   2. 26 miles
3. 45 hours

Practice, p. 38

1. Yes, 16 feet, 10 inches (or 16.806 ft.)
2. 75 bricks
3. 4
4. 225

Unit 3

Self-Check, pp. 42–43

1. proper
2. improper
3. complex
4. proper
5. mixed
6. mixed
7. improper
8. mixed

Practice, p. 43

1. complex   2. improper
3. proper   4. complex
5. improper   6. improper
7. complex

Self-Check, p. 44

1. $\frac{6}{18}$ 2. $\frac{16}{36}$ 3. $\frac{3}{7}$
4. $\frac{3}{4}$ 5. $\frac{21}{28}$ 6. $\frac{1}{6}$

Self-Check, p. 44

1. $\frac{1}{3}$ 2. $\frac{1}{3}$ 3. $\frac{1}{3}$
4. $\frac{1}{6}$ 5. $\frac{2}{5}$ 6. $\frac{2}{5}$

Self-Check, p. 45

1. 5   2. 60   3. 5
4. 2   5. 5   6. 20

7. 7   8. $3\frac{2}{5}$ 9. $2\frac{5}{13}$

Self-Check, p. 45

1. $\frac{107}{11}$ 2. $\frac{37}{3}$ 3. $\frac{436}{27}$
4. $\frac{87}{6}$ 5. $\frac{85}{4}$ 6. $\frac{297}{7}$
7. $\frac{118}{15}$ 8. $\frac{109}{22}$ 9. $\frac{34}{11}$

Practice, p. 45

1. $\frac{1}{4}$ 2. $\frac{3}{5}$ 3. $\frac{2}{3}$
4. $\frac{3}{5}$ 5. $1\frac{6}{11}$ 6. 8
7. $\frac{12}{17}$ 8. $\frac{5}{6}$ 9. $32\frac{5}{8}$
10. $\frac{140}{149}$ 11. $11\frac{4}{5}$ 12. 8
13. $\frac{5}{9}$ 14. $\frac{2}{9}$ 15. $2\frac{1}{15}$
16. $17\frac{3}{4}$ 17. $2\frac{5}{8}$ 18. 8
19. $12\frac{2}{9}$ 20. $\frac{1}{20}$ 21. $\frac{1}{3}$

Self-Check, p. 46

1. $\frac{4}{9}$ 2. $\frac{3}{4}$ 3. $\frac{9}{11}$
4. $\frac{2}{3}$ 5. $1\frac{2}{7}$ 6. $4\frac{1}{2}$
7. $10\frac{1}{5}$ 8. $9\frac{1}{6}$ 9. $4\frac{21}{25}$
10. $8\frac{9}{13}$

Practice, p. 47

1. $\frac{49}{52}$ 2. $1\frac{2}{5}$ 3. 10
4. $5\frac{5}{24}$ 5. $1\frac{11}{16}$

Self-Check, p. 47

1. $\frac{1}{2}$ 2. $\frac{2}{5}$ 3. $\frac{3}{11}$
4. $\frac{1}{4}$ 5. $\frac{1}{3}$ 6. $3\frac{4}{9}$
7. $2\frac{5}{7}$ 8. $3\frac{2}{3}$ 9. $5\frac{1}{5}$
10. $1\frac{1}{3}$

Self-Check, p. 48

1. $\frac{8}{15}$ 2. $\frac{3}{14}$ 3. $\frac{1}{10}$
4. $24\frac{11}{15}$ 5. $\frac{13}{20}$

Self-Check, p. 48

1. $11\frac{1}{4}$ 2. $1\frac{7}{9}$ 3. $4\frac{1}{15}$
4. $1\frac{4}{33}$ 5. $11\frac{1}{4}$

Self-Check, p. 49

1. $6\frac{2}{3}$ 2. $2\frac{3}{4}$ 3. $3\frac{1}{9}$
4. $\frac{7}{11}$ 5. $2\frac{2}{9}$ 6. $2\frac{1}{5}$
7. $17\frac{2}{3}$ 8. $4\frac{5}{6}$ 9. $3\frac{2}{3}$
10. $2\frac{23}{54}$

Self-Check, p. 49

1. $\frac{4}{21}$ 2. $\frac{5}{24}$ 3. $\frac{11}{27}$
4. $\frac{1}{9}$ 5. $\frac{11}{45}$ 6. $\frac{3}{7}$

7. $\frac{1}{15}$ 8. $\frac{7}{12}$ 9. $\frac{20}{27}$

Self-Check, p. 50

1. $\frac{25}{84}$ 2. $13\frac{2}{3}$ 3. $4\frac{7}{8}$
4. 21   5. $43\frac{3}{4}$ 6. $2\frac{16}{27}$
7. $28\frac{1}{9}$ 8. $\frac{7}{8}$ 9. $\frac{21}{32}$

Self-Check, p. 50

1. $1\frac{7}{8}$ 2. $1\frac{3}{10}$ 3. $2\frac{1}{7}$
4. $1\frac{32}{45}$ 5. $2\frac{2}{7}$ 6. $2\frac{5}{17}$
7. $1\frac{7}{8}$ 8. $\frac{25}{27}$ 9. 1

Self-Check, p. 51

1. $3\frac{4}{15}$ 2. $4\frac{1}{8}$ 3. $3\frac{3}{14}$
4. $6\frac{7}{10}$ 5. $5\frac{21}{25}$ 6. $\frac{35}{81}$

Practice, p. 51

1. $1\frac{1}{4}$ 2. $7\frac{3}{5}$ 3. 1
4. 26   5. $\frac{3}{8}$ 6. $5\frac{4}{15}$
7. $1\frac{35}{37}$ 8. $19\frac{2}{45}$ 9. $\frac{5}{8}$

Self-Check, p. 52

1. .24   2. .8   3. .875
4. .9   5. .4   6. .4

Self-Check, p. 52

1. .83   2. .92   3. 7.67
4. 4.6   5. 1.2   6. 9.13

Practice, p. 52

1. .38   2. .90   3. 1.17
4. .53   5. .42   6. 3.33
7. 15.2   8. .67   9. 2.71

Self-Check, p. 54

1. $\frac{1}{20}$ 2. $16\frac{1}{2500}$ 3. $2\frac{69}{100}$
4. $\frac{23}{80}$ 5. $\frac{9}{10}$ 6. $1\frac{13}{200}$
7. $\frac{7}{40}$ 8. $\frac{1}{4}$ 9. $\frac{73}{100}$
10. $\frac{111}{125}$

Practice, p. 54

1. $\frac{37}{100}$ 2. $4\frac{64}{125}$ 3. $2\frac{1}{2}$
4. $11\frac{11}{40}$ 5. $9\frac{327}{500}$ 6. $11\frac{1}{20}$
7. $\frac{311}{1,000}$ 8. $3\frac{303}{500}$ 9. $24\frac{69}{200}$

Unit 4

Self-Check, p. 58

1. 1%   2. 12%   3. 5%
4. 69%   5. 26%   6. 1%

Practice, p. 58

1. 23%   2. 40%   3. 75%

4. 15%   **5.** 100%   **6.** 25%

Self-Check, p. 59

1. 40.4%   **2.** 20%   **3.** 34.2%
4. 6%   **5.** 333%   **6.** 9%
7. .5%   **8.** 12%   **9.** 3,433.2%

Self-Check, p. 59

1. .56   **2.** .35   **3.** .21
4. .1507   **5.** .999   **6.** .752
7. .3333   **8.** .50   **9.** .0009

Self-Check, p. 60

1. 60%   **2.** 44%   **3.** 25%
4. 50%   **5.** 20%   **6.** 12.5%

Self-Check, p. 60

1. $\frac{1}{2}$ **2.** $\frac{17}{50}$ **3.** $\frac{13}{20}$
4. $\frac{87}{200}$ **5.** $\frac{13}{125}$ **6.** $\frac{3}{4}$
7. $\frac{189}{250}$ **8.** $\frac{111}{200}$ **9.** $\frac{111}{1,000}$

Practice, p. 61

1. .6%   **2.** 15%
3. 175.5%   **4.** 67%
5. 300.1%   **6.** 75.63%
7. 675%   **8.** 5.07%
9. 375%   **10.** .175
11. 1.152   **12.** .5
13. .079   **14.** 3.1765
15. .0045   **16.** .54
17. 5.13   **18.** .009
19. 77.78%   **20.** 42.86%
21. 45.45%   **22.** 87.5%
23. 50%   **24.** 33.33%
25. $\frac{1}{5}$ **26.** $\frac{1}{2}$
27. $\frac{1}{10}$ **28.** $\frac{3}{4}$
29. $\frac{3}{5}$ **30.** $\frac{33}{100}$
31. $\frac{97}{200}$ **32.** $\frac{143}{500}$
33. $\frac{441}{500}$

Self-Check, p. 62

1. $77.4   **2.** $1.14   **3.** 1.05
4. .28   **5.** 854   **6.** .15

Self-Check, pp. 62–63

1. 40%   **2.** 40%   **3.** 31.25%
4. 60%   **5.** 420   **6.** 525
7. $425   **8.** 1,700

Practice, p. 63

1. $187.50   **2.** $.0385
3. $1.995   **4.** 923
5. 780   **6.** 32
7. 15%   **8.** 5.5%
9. 5%   **10.** 100%
11. 20%   **12.** 250%
13. 3,000   **14.** 50
15. 502   **16.** 2,750

17. 300   **18.** 160

Self-Check, p. 64

1. 1 : 2   **2.** 7 to 1
3. 1 : 3   **4.** 10 to 1
5. 50 : 1   **6.** 100 to 1

Self-Check, p. 65

1. $2\frac{1}{2}$ miles per hour
2. 150 miles per hour
3. $.35 for photo
4. 6.7 miles per hour
5. $3.85 per hour

Practice, pp. 65–66

1. 70 to 120 = 7 : 12 (7 to 12)
2. 300 to 220 = 15 to 11
(15 : 11)
3. 300 to 240 = 5 to 4 (5 : 4)
4. 120 to 240 = 1 : 2 (1 to 2)
5. 70 to 240 = 7 : 24 (7 to 24)
6. 50 miles per hour
7. $.54 per pound
8. 7.5% per ounce
9. $.43 per pound
10. $.218 per pound
11. 1.98
12. 10
13. 115
14. $1,000
15. $12\frac{1}{2}$%

Unit 5

Self-Check, p. 70

1. $116.40   **2.** $79.20
3. $230.63   **4.** $73.94
5. $232.50

Self-Check, pp. 71–72

1. $329   **2.** $570
3. $192.38   **4.** $465.45
5. $294.30

Self-Check, p. 72

1. $550   **2.** $280   **3.** $315
4. $55   **5.** $700   **6.** $900

Self-Check, p. 73

1. $31,200   **2.** $576.92
3. $10,416.67   **4.** $36,000
5. $5,200   **6.** $775.83

Self-Check, p. 74

1. $680.80   **2.** $1,200

3. $1,548   **4.** $985.50

Self-Check, p. 75

1. $2,704.80   **2.** $1,928
3. $2,300   **4.** $860, $860

Self-Check, p. 76

1. $1,700   **2.** $975

Practice, pp. 76–77

1. $96   **2.** $112.50
3. $206.25   **4.** $1,583.33
5. $40.75   **6.** $17,745
7. $3,025   **8.** $350

Self-Check, p. 80

1. $380—$39   **2.** $240—$29
3. $279.30—$27   **4.** $327.45—$32
5. $384.80—$32

Self-Check, p. 81

1. $23.58   **2.** $187.75
3. $180.24   **4.** $3,379.50
5. $35.75

Self-Check, pp. 82–83

1. $186.19   **2.** $11,165.60
3. $277.91   **4.** $10,761.13
5. $10,255.87

Practice, pp. 83–84

1. $381.15
2. Alpha; $18,000.32; $17,492.80
3. $316.35
4. $244.59

Unit 6

Self-Check, p. 89

Deposit slip
bills   120.00
coins   5.50
checks:
42–220   49.50
21–874   213.78
21–356   49.33
Total   438.11

Practice, p. 89

1. $400.91   **2.** $1,054.83
3. $134.00   **4.** $590.56
5. $809.21

Self-Check, pp. 90–91

Check No. 111; Current date;
Pay to the order of:
Welch Auto Repair; $111.00;
One hundred eleven and $\frac{00}{100}$;
repairs, none

Self-Check, p. 92

Checkbook Register
$764.59, $723.24, $523.24

Practice, p. 92

Checkbook Register
ending balance: $472.73

Self-Check, p. 95

	New Register Balance	Adjusted Statement Balance
1.	$821.25	no changes
2.	$423.47	$423.27
3.	$340.81	$340.81
4.	$946.47	$946.47
5.	$60.06	$ 60.06

Practice, p. 95

1. $939.50; $939.50
2. a. $70.80 b. $84 c. $57
3. Reconciled balances = $552.44

Unit 7

Self-Check, p. 100

1. $205.45   2. $8,500
3. $197.56   4. $973.27

Self-Check, p. 101

1. $73.13   2. $212.50
3. $1,124.22

Self-Check, p. 102

1. $300   2. 8%
3. 2 years

Practice, pp. 102–103

1. $48.13   2. $96.58
3. $61.11   4. $34.50
5. 8%

Self-Check, p. 106

1. $28.07   2. $34.39
3. $261.96   4. $298.01
5. $1,882.50

Self-Check, p. 107

1. $133.39   2. $5,435
3. $11,282.85

Self-Check, p. 108

1. $5,422.97   2. $151
3. $502.72   4. $14,307.21

Practice, p. 108

1. $376.50   2. $76.12

3. $1,346.73   4. $2,015.57
5. $54.69

Self-Check, p. 109

1. $45   2. $36.16
3. $555.56

Self-Check, p. 110

	Interest	Maturity Value
1.	$84.16	$1,684.16
2.	$86.30	$3,586.30
3.	$18.86	$887.86
4.	$71.54	$5,346.54
5.	$189.74	$2,889.74

Self-Check, p. 111

	Bank Discount	Proceeds
1.	$63.89	$1,776.11
2.	$53.33	$1,946.67
3.	$397.50	$4,102.50
4.	$5.42	$644.58
5.	$245.00	$6,755.00

Practice, pp. 111–112

1. $1,808.22   2. $4,110
3. $18,438   4. $9.86
5. $5,123.29

Unit 8

Self-Check, p. 118

1. $245.28   2. 15%
3. 18.5%

Practice, p. 118

1. 16.75%
2. $892.80, $725.40, $167.40
3. $297.12
4. $81.76
5. $148.40

Self-Check, p. 119

	Tax	Total
1.	$1.65	$34.64
2.	$1.50	$31.50
3.	$2.45	$51.45
4.	$1.13	$23.63
5.	$1.70	$35.70

Self-Check, p. 121

1. $700   2. $199.32
3. $406.96

Self-Check, p. 121

1. $800   2. $2,000
3. $755   4. $1,144
5. $262   6. $1,789

7. $978

Practice, p. 122

1. 0
2. $923.59
3. $807.54, $40.38, $847.92

Self-Check, p. 123

1. 20.4% 2. 1.37% 3. 1.05%

Self-Check, pp. 123–124

1. $2.69 2. $20.08 3. $4.01

Self-Check, p. 124

1. $2.11   2. $17.01

Self-Check, p. 125

1. $233.33   2. $193.46

Practice, pp. 125–126

1. $28.20

	Unpaid Balance	Finance Charge
2.	$129.70	$1.69
3.	$46.69	$.56
4.	$662.98	$13.26

Unit 9

Self-Check, p. 130

1. $7,530.75   2. $42.50
3. $32

Self-Check, p. 131

1. $269.55   2. $300

Self-Check, pp. 131–132

1. 25% 2. 10% 3. $27\frac{1}{4}$%

Practice, p. 132

1. $132   2. $150
3.

	Trade Discount	Trade Discount Rate
	$9.60	20
	$4.27	12
	$4.49	15

Self-Check, pp. 134–135

1. $921.50   2. $1,543.30
3. $3,758.75   4. $752.64
5. June 25: $419.77
   July 5: $424.06

Practice, p. 135

1. $2,594.70   2. $846.94
3. $545.96   4. $3,058.47

5. $2,690.00

Self-Check, p. 137

1. $319.59
2. <u>Net Price</u>
   $42.81
   $38.00
   $51.32

Practice, p. 137

1. $151.64   2. $410.40
3. $36.30

Self-Check, p. 138

1. $.79   2. $.43   3. $2,200

Self-Check, p. 139

	Markup	Markup Rate
1.	$53	67%
2.	$3,160	55%
3.	$24.72	25%

Self-Check, p. 139

	Markup	Markup Rate
1.	$.20	15%
2.	$.34	97%
3.	$.83	16%
4.	$.39	15%
5.	$.10	34%

Practice, p. 140

	Markup	Markup Rate
1.	$21.45	35%
2.	$3.64	25%
3.	$80.50	75%
4.	$.20	91%
5.	$1.30	12%
6.	$.23	37%

Index